HEALTH SCIENCES LIBRARY
D0295174

# DIAGNOSIS

# DIAGNOSIS

## DISPATCHES FROM THE FRONTLINES OF MEDICAL MYSTERIES

LISA SANDERS

ICON BOOKS

Originally published by Broadway Books in 2009 with the title *Every Patient Tells a Story: Medical Mysteries and the Art of Diagnosis*

Published by arrangement with Broadway Books, part of The Crown Publishing Group, a division of Random House, Inc.

Some of the material in this book appeared in different form in the *New York Times*.

Published in the UK in 2009 by
Icon Books Ltd, Omnibus Business Centre,
39–41 North Road, London N7 9DP
email: info@iconbooks.co.uk
www.iconbooks.co.uk

Sold in the UK, Europe, South Africa and Asia
by Faber & Faber Ltd, Bloomsbury House,
74–77 Great Russell Street, London WC1B 3DA

Distributed in the UK, Europe, South Africa and Asia
by TBS Ltd, TBS Distribution Centre, Colchester Road,
Frating Green, Colchester CO7 7DW

ISBN: 978-184831-072-8



Typeset in Adobe Garamond by Marie Doherty

Printed and bound in the UK by
CPI Mackays, Chatham, ME5 8TD

*To Jack*

# *CONTENTS*

## AUTHOR'S NOTE

The stories I tell here are real. In order to respect the confidentiality of these patients who were kind enough to share their stories with me, I have changed their names. In some instances I have altered certain identifying details as well. The doctors featured in these pages described in detail some of their most difficult diagnoses – mistakes and all. They are distinguished not by these errors but by their willingness to discuss them. No one should be punished for simply being honest, and so I have changed the names of these brave doctors.

The use of pronouns when you are speaking of an individual remains problematic in writing now that we can no longer just use the generic 'he.' There is no rule on this at this point, so in this book I will refer to the generic doctor as she and the generic singular patient as he.

# INTRODUCTION

## *Every Patient's Nightmare*

Barbara Lessing stared out the window at the snowy field behind the hospital. The afternoon sky was dark with yet more snow to come. She looked at the slender figure in the bed. Her daughter, Crystal, barely twenty-two years old and healthy her entire life, was now – somehow – dying. The young woman had been in the Nassau University Medical Center ICU for two days; she'd been seen by a dozen doctors and had scores of tests, yet no one seemed to have the slightest idea of just what was killing her.

It all started at the dentist's office. Crystal had had a couple of impacted wisdom teeth taken out the month before. But even after the teeth were gone, the pain persisted. She'd called her mother halfway across the state just about every day to complain. 'Call your dentist,' she'd urged her daughter. And she had. Finally.

The dentist gave her a week's worth of antibiotics and then another. After that her mouth felt better – but she didn't. She was tired. Achy. For the next week she'd felt like she was coming down with something. Then the bloody diarrhoea started. And then the fevers. Why didn't you go to the doctor sooner? the trim middle-aged woman scolded her daughter silently.

Barbara had had a call from a doctor in the emergency room of this suburban hospital the night before. Her daughter was ill, he told her. Deathly ill. She drove to Syracuse, caught the next flight to New York City, and drove to the sprawling academic medical centre on Long Island. In the ICU, Dr Daniel Wagoner, a resident in his second year of training, ushered her in to see her daughter. Crystal was asleep, her dark curly hair a tangled mat on the pillow. And she looked very thin. But most terrifying of all – she was yellow. Highlighter yellow.

Wagoner could feel his heart racing as he stood looking at this jaundiced wisp of a girl lying motionless on the bed. The bright unnatural yellow of her skin was shiny with sweat. She had a fever of nearly 103°. Her pulse was rapid but barely palpable and she was breathing much faster than normal despite the oxygen piped into her nose. She slept most of the time now and when awake she was often confused about where she was and how she had come to be there.

To a doctor, nothing is more terrifying than a patient who is dying before your eyes. Death is part of the regular routine of the ICU. It can be a welcome relief to the patient, or to his family. Even a doctor may accept it for a patient whose life can be prolonged no longer. But not for a young girl who was healthy just weeks ago. These doctors had done everything they could think of but still there was a fear – a reasonable fear – that they'd missed some clue that could mean the difference between life and death for this young woman. She shouldn't die, but the young resident and all the doctors caring for her knew that she might.

Crystal's thin chart was filled with numbers that testified to how very ill she was. Wagoner had been through the chart a dozen times. Virtually every test they'd run was abnormal. Her white blood cell count was very high, suggesting an infection. And her red blood cell count was low – she had barely half the amount of blood she should have. She'd had a transfusion in the emergency room and another after she was moved to the ICU, but her blood count never budged. Her kidneys weren't working. Her clotting system wasn't either. Her yellow skin was covered in bruises and her urine was stained deep red.

Sometimes, if you just work hard enough to keep a patient alive – to keep the blood circulating, the lungs oxygenating, the blood pressure high enough – the body will be able to survive even a vicious illness. These are the miracles brought by technological advances. Sometimes, but not this time. The ICU team gave Crystal bag after bag of blood; they did their best to shore up her damaged clotting system; she got pressers (medications designed to increase blood pressure) and fluids to help her kidneys. She was on several broad-spectrum antibiotics. And yet none of that was enough. She needed a diagnosis. Indeed, she was dying for a diagnosis.

This book is about the process of making that diagnosis, making any diagnosis. So often this crucial linchpin of medicine goes unnoticed and undescribed, yet it is often the most difficult and most important component of what physicians do. As pervasive as medicine has become in modern life, this process remains mostly hidden, often misunderstood, and sometimes mistrusted. In movies and novels it's usually the one-liner that separates the fascinating symptoms from the initiation of the life-saving therapy. On television it's the contemporary version of Dr McCoy's (*Star Trek*) magic diagnostic device (his tricorder) that sees all, tells all. But in real life, the story of making a diagnosis is the most complex and exciting story that doctors tell.

And these are stories that doctors tell. Just as Sherlock Holmes or Nick Charles (the hero of the *Thin Man* mysteries) or Gil Grissom (*CSI*) delights in explaining the crime to victims and colleagues, doctors take pleasure in recounting the completed story of their complex diagnoses, stories where every strange symptom and unexpected finding, every mystifying twist and nearly overlooked clue, finally fit together just right and the diagnosis is revealed. In this book I'll take you into those conversations and onto the front lines where these modern medical mysteries are solved – or sometimes not.

Just a hundred years ago, journalist and acerbic social critic Ambrose Bierce defined the word 'diagnosis' in his *Devil's Dictionary* as 'A physician's forecast of disease by [taking] the patient's pulse and purse.' And that was true for most of human history. Until very recently, diagnosis was much more art than science.

But since Ambrose Bierce wielded his rapier pen, there has been a revolution in our ability to identify the cause of symptoms and understand the pathology behind them. In the era in which Bierce wrote, Sir William Osler, considered by many to be the father of American medicine, was able to write a comprehensive summary of all the known diseases in his 1,100-page masterwork, *The Principles and Practice of Medicine*. These days each tiny sub-branch of medicine could provide as many pages on its super-specialised knowledge alone.

At the birth of medicine, millennia ago, diagnosis (the identification of the patient's disease) and prognosis (the understanding of the disease's likely course and outcome) were the most effective tools a doctor brought to the patient's bedside. But beyond that, little could be done to either confirm a diagnosis or alter the course of the disease. Because of this impotence in the face of illness, the consequences of an incorrect diagnosis were minimal. The true cause of the illness was often buried with the patient.

In more recent history, medicine has developed technologies that have transformed our ability to identify and then treat disease. The physical exam – invented primarily in the nineteenth century – was the starting point. The indirect evidence provided by touching, listening to, and seeing the body hinted at the disease hidden under the skin. Then the X-ray, developed at the start of the twentieth century, gave doctors the power to see what they had previously only imagined. That first look through the skin, into the inner structures of the living body, laid the groundwork for the computerised axial tomography (CT) scan in the 1970s and magnetic resonance imaging (MRI) in the 1990s. Blood tests have exploded in number and accuracy, providing doctors with tools to help make a definitive diagnosis in an entire alphabet of diseases from anaemias to zoonoses.

Better diagnosis led to better therapies. For centuries, physicians had little more than compassion with which to help patients through their illnesses. The development of the randomised controlled trial and other statistical tools made it possible to distinguish between therapies that worked and those that had little to offer beyond the body's own recuperative powers. Medicine entered the twenty-first century stocked with a pharmacopoeia of potent and effective tools to treat a broad range of diseases.

Much of the research of the past few decades has examined which therapies to use and how to use them. Which medication, what dose, for how long? Which procedure? What's the benefit? These are all questions commonly asked and that can now be regularly and reliably answered. Treatment guidelines for many diseases are published, available, and regularly used. And despite concerns and lamentations about 'cookbook medicine,' these guidelines, based on a rapidly growing foundation of evidence, have saved lives. These forms of evidence-based medicine allow patients to benefit from the thoughtful application of what's been shown to be the most effective therapy.

But effective therapy depends on accurate diagnosis. We now have at our disposal a wide range of tools – new and old – with which we might now make a timely and accurate diagnosis. And as treatment becomes more standardised, the most complex and important decision making will take place at the level of the diagnosis.

Often the diagnosis is straightforward. The patient's story and exam suggest a likely suspect and the technology of diagnosis rapidly confirms the hunch. An elderly man with a fever and a cough has an X-ray revealing a raging pneumonia. A man in his fifties has chest pain that radiates down his left arm and up to his jaw, and an EKG (electrocardiogram) or blood test bears out the suspicion that he is having a heart attack. A teenage girl on the birth control pill comes in complaining of shortness of breath and a swollen leg, and a CT scan proves the presence of a massive pulmonary embolus. This is the bread and butter of medical diagnosis – cases where cause and effect tie neatly together and the doctor can almost immediately explain to patient and family whodunit, how, and sometimes even why.

But then there are the other cases: patients with complicated stories or medical histories; cases where the symptoms are less suggestive, the physical exam unrevealing, the tests misleading. Cases in which the narrative of disease strays off the expected path, where the usual suspects all seem to have alibis, and the diagnosis is elusive. For these, the doctor must don her deerstalker cap and unravel the mystery. It is in these instances where medicine can rise once again to the level of an art and the doctor-detective must pick apart the tangled strands of illness, understand which questions to ask, recognise the subtle physical findings, and identify which tests might lead, finally, to the right diagnosis.

To the doctors caring for Crystal Lessing, it was not clear if the mystery of her illness was going to be solved in time to save her life. Certainly there was no shortage of diagnostic data. There were so many abnormalities it was difficult to distinguish between the primary disease process and those that were the downstream consequences of the disease. The doctors in the ER had focused on her uncontrolled bleeding. Why wasn't her blood clotting? Was this disseminated intravascular coagulopathy (DIC) – a mysterious disorder that frequently accompanies the most severe infections? In this disease the fibrous strands that make up a clot form willy-nilly inside blood vessels. These tough strands slice through red blood cells as they course through the artery, releasing the oxygen-carrying contents and strewing the torn fragments of cells into the circulation. Yet careful examination of Crystal's blood didn't reveal any of these cell membrane fragments. So it wasn't DIC.

And why was she yellow? Hepatitis was the most common cause of jaundice in a young person. But the ER physician found no evidence of any of the several viruses that can cause hepatitis. Besides, the blood tests they'd sent to check how well her liver was working were almost normal. And so, they concluded, it wasn't her liver.

Once Crystal was transferred to the ICU, the doctors there had focused on the bloody diarrhoea. She'd had two courses of powerful antibiotics for a dental infection before the diarrhoea and fever started. That fit the pattern for an increasingly common infection with a bacterium called *Clostridium difficile*, or *C. diff.*, as it's known around the hospital. The use of antibiotics can set the stage for this bacterial infection of the colon, which causes devastating diarrhoea and a severe, sometimes fatal, systemic illness.

The ICU team had looked for the dangerous toxin made by the *C. diff.* bacteria but hadn't found it. Still, that test can miss up to 10 per cent of these infections. In fact, it's standard practice to retest for the bacterial toxin three times before believing that the disease isn't present when suspicion for the disease is high. The ICU team started Crystal on antibiotics to treat *C. diff.* anyway – the story of antibiotics followed by bloody diarrhoea made that their leading diagnosis.

But Dr Wagoner, the resident caring for the patient, was unsatisfied with the diagnosis. Too many pieces didn't seem to fit. The antibiotics and diarrhoea made sense but the diagnosis left too many of her symptoms unexplained.

That Friday afternoon – forty-eight hours after Crystal was admitted to the hospital – Wagoner did what doctors often do when faced with a complex case: he reached out to a more experienced physician. Despite all the available technology, the tools doctors often rely on most are the most old-fashioned – a phone, a respected colleague, a mentor or friend.

Dr Tom Manis was one of the most highly regarded doctors in the hospital. A nephrologist, he was called in because of Crystal's kidney failure. But as Wagoner presented the patient to the older doctor, it was clear he was hoping that Manis could help them figure out more than just the kidney.

As Manis read through the chart, he too became alarmed. Wagoner was right – this diagnosis didn't fit well at all. For one thing, *C. diff.* colitis is usually a disease of the sick and elderly. The patient was young and had been healthy. But even more to the point, *C. diff.* wouldn't account for the profound jaundice and the anaemia that persisted despite multiple transfusions. So Manis did what the resident had done – 'I called every smart doctor I knew,' and told them each the perplexing tale of Crystal Lessing – again, using those irreplaceable tools, a phone and a friend. One of those friends was Dr Steven Walerstein, the head of the hospital's Department of Medicine.

It was early evening by the time Walerstein had a chance to see the patient. He didn't read her chart. He never did in tough cases like this. He didn't want to be influenced by the thinking of those who had already seen her. Far too often in these difficult cases something has been missed, or misinterpreted. And even if they had collected all the pieces, they had clearly put the story of this illness together incorrectly.

Instead he went directly to the patient's bedside.

Walerstein introduced himself to the young woman and her mother. He pulled up a chair and sat down. Getting the whole story is essential but it can take time. Can you tell me what happened, from the beginning? he asked the sick girl. Like the classic detective in a mystery novel, he asked the victim to go over the crime once more. 'I've told this story so many times,' Crystal protested. Her voice was thick with fatigue, her words slurred. Couldn't he just read it in her chart? No, he told her gently but firmly. He needed to hear it from her, needed to put it together for himself. Slowly the girl began telling her story once more. Her mother took up the tale when the girl became confused or couldn't remember.

Once the two women had gone through the events that brought each of them to the ICU, Walerstein asked the mother for a little more information about her daughter. Crystal had just graduated from college, she told him. She was working as a nanny while she tried to figure out what she wanted to do with her life. She didn't smoke or drink or use drugs. And she'd never been sick. Never. She roughly brushed away tears as she described her daughter to this kindly middle-aged doctor. He nodded sympathetically. He had a daughter.

Then Walerstein turned back to the young woman in the bed. Her yellow skin was now hot and dry. Her lips were parched and cracked. Her abdomen was distended and soft, but he could feel the firm edge of the liver, normally hidden by the rib cage, protruding a couple of inches below. She moaned again as he put pressure on this tender and enlarged organ.

Only then did he allow himself to look through her chart. He skipped over the notes and buried himself in the myriad abnormal test results that had been collected over her two days in the intensive care unit.

Walerstein was a general internist (doctor of internal medicine), admired for his broad knowledge of medicine and his clinical acumen. If he didn't know the answer right off the bat, he was known to ask questions that would lead to the answer. And this young woman needed an answer, or she would die. Having thoroughly examined the patient and her chart, Walerstein took a moment to step back and look for some kind of pattern buried in the chaos of numbers and tests.

The ICU doctors had focused on the bloody diarrhoea and had got nowhere. Indeed, although the girl had seen blood in her stools at home, since arriving at the hospital she had very little diarrhoea. It didn't seem to Walerstein to be the most important of her symptoms. Instead, Walerstein went back to the striking feature that had caught the ER doctor's eye – her blood would not clot.

The liver makes most of the proteins that cause blood to coagulate. Could it be that her liver was no longer making these proteins? *Could it be that her liver wasn't working at all?* That would account for both the bleeding and the jaundice. But liver failure is usually marked by dramatic elevations in certain enzymes that are released when liver cells are destroyed, and those enzymes had been nearly normal since she'd come to the hospital. Her doctors had taken that to mean that the liver was not involved in this deadly process.

What if, instead, the liver had already been destroyed by the time Crystal came to the hospital? What if these markers of liver injury (known as transaminases) weren't elevated because there were no more liver cells left to injure, if all the liver cells had already been destroyed? No one in the Emergency Department or in the ICU had made this leap. And yet if you looked at it in this way, as Walerstein did, everything made perfect sense. It all fit.

He then turned his attention to the profound anaemia that had been noted from the start. Despite multiple transfusions, Crystal still had only half as much blood as she should. She was bleeding – her red-tinted urine showed that – but she wasn't bleeding that much. It was clear that her red blood cells were being destroyed within her body. Deep within her chart there was a test that showed this but Walerstein noted that the team caring for her hadn't considered this in their search for a diagnosis.

Too often information you don't initially understand is simply set aside, especially when there is such a wealth of information. Walerstein understood this phenomenon. And once set aside it's often forgotten. It happens all the time. But Walerstein also knew that in a difficult case like this one, data that has been set aside often holds the key.

So Crystal had liver failure and red blood cell destruction. That combination stirred something deep within his memory. Walerstein could feel the pieces slowly come together like the cogs in some ancient machine. And then suddenly he knew what this was.

The internist hurried to the library to check his hunch. Yes! He was right. This combination – liver failure and red blood cell destruction – was an unusual manifestation of an unusual inherited illness: Wilson's disease.

In Wilson's disease, the liver lacks the machinery to regulate copper, an essential mineral found in the diet. Without these chemical tools, excess copper builds up in the liver and other organs and slowly, insidiously breaks them down. Usually this process takes place over decades, but occasionally, for reasons that are still not understood (though it is often associated with the use of antibiotics, as it was in Crystal's case), the copper blasts out of the liver – destroying the organ in the process – and a lifetime of the stored mineral floods into the bloodstream. Once there, all hell breaks loose: the copper demolishes red blood cells on contact. The kidneys work hard to clear the cell fragments from the circulation but are gravely injured in the process. Meanwhile the high levels of copper in the bloodstream attack virtually every organ in the body. In this form, the disease is rapidly and universally fatal unless the patient receives the only possible cure – a new liver to replace the one destroyed by the jailbreak of copper, a liver that has the machinery to dispose of the excess mineral. If this was Wilson's disease, this patient needed a transplant immediately.

But first Walerstein had to confirm the diagnosis. It was late on a Friday night and so it would be impossible to measure the amount of copper in her blood – in any case, his hospital laboratory didn't even do that test. But there was another way to diagnose this disease. Patients with Wilson's will often accumulate copper in their eyes – a golden brown ring at the very outermost edge of the iris. Walerstein hurried back to the ICU. He carefully examined the girl's eyes. Nothing. He couldn't see the rings, but maybe an ophthalmologist with his specialised equipment could. 'It's not often that you call the ophthalmologist at nine p.m. on a Friday' to do an emergency examination, Walerstein told me. But he related the girl's story one more time – this time with a likely diagnosis, if only he could confirm it. 'I'm sure he thought I was nuts, until he saw the rings.' As soon as Walerstein had the results, he hurried into the patient's room to tell the girl and her mother what they'd found.

Crystal Lessing was transferred by helicopter that night to New York–Presbyterian Hospital. Patients with the greatest need get priority in the transplant line. Without a new liver, Crystal would die within days and that put her at the front of the line. She received an organ the following week and survived.

Crystal's story is every patient's nightmare: To be sick, even dying, and have doctor after doctor fail to figure out why. To be given the wrong diagnosis, or no diagnosis at all, and to be left to the ravages of disease with nothing more than your own endurance and the doctors' best-guess therapy to rely on. To live or die in a modern hospital filled with the promise of treatment and yet without a diagnosis to guide its use.

How was Walerstein finally able to make a diagnosis after so many others had failed? How do doctors make these tough diagnoses? Walerstein is modest about his role in the case. 'I think I was just lucky enough to know about this rare form of this rare disease. No one can know everything in medicine. I happened to have known about this,' he told me. It's sometimes a mysterious process – even to the doctors themselves. 'A bell went off and the connection was made,' Walerstein told me. 'That's all I know.'

This book is about that bell – how doctors know what they know and how they apply what they know to the flesh-and-blood patient who lies before them. It can be a messy process, filled with red herrings, false leads, and dead ends. An important clue may be overlooked in the patient's history or examination. An unfamiliar lab finding may obscure rather than reveal. Or the doctor may be too busy or too tired to think through the case. Even the great William Osler must have had his bad days.

And the patient, by definition sick, frequently tired and in pain, inarticulate with distress, is given the essential task of telling the story that could help a doctor save his life. It's a recipe for error and uncertainty. It is 'an inferential process, carried out under conditions of uncertainty, often with incomplete and sometimes inconsistent information,' says Jerome Kassirer, former editor of the *New England Journal of Medicine* and one of the earliest and most thoughtful modern writers on this unruly process.

It's a wayward path to an answer filled with unreliable narrators – both human and technological – and yet, despite the unlikeliness, that answer is often reached and lives are saved.

Often, but not always. The possibility of error is ever present.

It's certainly not news that medical errors are common. In 1999 the National Institutes of Health (NIH), Institute of Medicine, released a report on the topic – *To Err Is Human*. In that now famous report the authors concluded that there were up to 98,000 patient deaths due to medical errors every year – the same number of deaths we would see if a jumbo jet crashed every day for a year. That set off a national effort to reduce the rate of errors in medicine that is still bearing fruit.

That report did not look at errors in diagnosis. And yet errors in diagnosis make up a large chunk of the errors made in medicine. Depending on which study you believe, it is the first or second most common cause of medical lawsuits. Studies suggest that between 10 and 15 per cent of patients seen in primary care specialties – internal medicine, family medicine, and paediatrics – are given an incorrect diagnosis. Often the error has no effect – people get better on their own or return to their doctor when the symptoms get worse – but doctors and patients alike worry about the possibility of a diagnostic error that hurts or even kills. In a study of over thirty thousand patient records, researchers found that diagnostic errors accounted for 17 per cent of adverse events.

Doctors are getting better at making diagnoses. Tests and imaging have made possible diagnoses that were in earlier times only knowable at autopsy. And while postmortem studies done in the US suggest that the rate of unsuspected diagnoses has been remarkably stable over the past several decades, that statistic is skewed by the diminishing number of autopsies performed. A study done at the University Hospital in Zurich, Switzerland, where there is a 90 per cent autopsy rate, shows that over the past few decades the number of missed or erroneous diagnoses has steadily dropped. Another study done for the Agency for Healthcare Research and Quality (the AHRQ, a research arm of the NIH) shows a similar trend in the US if you account for the ever shrinking pool of autopsies.

Still, the fear of getting it wrong is always present for both doctors and patients. As a result, there is a new and growing interest in better understanding diagnostic errors in medicine. The first-ever conference on the topic – one of the earliest signs of growing research interest – was held in Phoenix in 2008. And the AHRQ, the government agency charged with improving the quality of health care in the US, offered its first grants for research on the topic in the autumn of 2007.

Research into diagnostic error, like research into the diagnosis process itself, is still a very new field. There is even difficulty in deciding what constitutes a diagnostic error. What a thoughtful patient may consider an error is not necessarily the same as that which his equally thoughtful doctor might consider an error.

For example, when a patient comes to my office with a sore throat and a fever, I might check for strep, and if it's not present I'll probably send him out with a diagnosis of a viral illness. But I share with all such patients what I expect to happen over the next few days – that they should start to feel better within a day or two. And if not, I tell them to call me and let me know. Because, while the odds are overwhelming that this is simply a viral syndrome, it's not 100 per cent certain. I might be wrong. The test might be wrong. It might be mono (glandular fever). It might be some other kind of bacterial tonsillitis. It might be cancer.

I can't just check under the hood and see if the spark plugs need to be replaced – the way a mechanic diagnoses the funny noise your car is making. Instead, I have to listen to the engine and, based on the indirect evidence I can collect, make a thoughtful and well-informed guess as to what is *probably* going on.

If I send that patient home with a diagnosis of a viral syndrome and he doesn't get better and has to come back, would that be a diagnostic error? I suspect the patient would think so. And certainly it wasn't a correct diagnosis. But did I make an error? Should I have done something different?

I could have been more certain. I could have sent my patient to an Ear, Nose, and Throat specialist who could have looked down his throat with a special scope. I could have even asked for a biopsy of the red and swollen tissue to confirm my diagnosis. That would be time-consuming and painful for the patient and ridiculously expensive. But even then, the diagnosis would not have been 100 per cent certain. In medicine, uncertainty is the water we swim in.

The chance of being wrong is overwhelming when dealing with something more complicated than a sore throat. Doctors – far more than the patients they care for – recognise that some error is inevitable. From the first moment a doctor sets eyes on a patient, she begins to formulate a list of possible causes of the symptoms – what's known as a differential diagnosis. As the story emerges, that list is modified – diseases on the list disappear to be replaced by new ones that more closely adhere to the patient's story, or exam, or sometimes test results. By the end of the encounter the doctor has a list of likely suspects.

If the doctor has worked through the problem well, there's a very good chance that one of these possible diagnoses will be right. The rest though, by definition, will be wrong. We are regularly wrong in the pursuit of being right. It's important to have a list of possibilities because medicine is complicated and diseases and bodies differ. We frequently have a diagnosis that we consider most likely, but we're taught also to come up with a plan B because our patients don't always have the most likely disease. The question we are taught to ask ourselves is, if it isn't that, what else could it be?

As a collector of diagnostic stories, I find myself frequently asking why one doctor was able to make the diagnosis when others before her couldn't. Where were the errors? How were they made? What can we learn?

Sometimes the problem is a lack of knowledge. This was certainly the case in Crystal Lessing's story. She had a rare presentation of an unusual disease. One of the human limitations in medicine is that no one can know everything.

There were errors in thinking in Crystal's case as well. Recognising that the fundamental problem was liver failure was an essential step in Walerstein's thinking process – a realisation that eluded all of the doctors who had seen her initially.

There were also mistakes in some of the data collected from the patient. Walerstein noted that the patient's 'bloody diarrhoea' consisted of a couple of episodes of bloody stools the day she came to the hospital. And Walerstein was also the first to note the enlarged and tender liver when he examined the young woman – a hint that the organ wasn't as normal as the blood tests suggested. Abnormalities uncovered by testing were also not interpreted correctly. Crystal's jaundice was initially attributed to the destruction of the red blood cells. Yet when further testing revealed that this red-blood-cell massacre was not the result of an abnormal immune system improperly attacking the cells, Walerstein was the first to consider other causes of red cell destruction. Research suggests that diagnostic errors – like this one nearly was – are often due to a multitude of missteps made along the way.

The solution to this case, as with so many cases, lay in the proper use of all the tools we have at our disposal. Walerstein took a careful history, performed a thorough physical examination, and identified the important lab abnormalities. Only then was he able to connect the information about that patient with the knowledge he had to make the diagnosis. Only then did the pieces of the puzzle come together.

In telling you these stories I try to put you, the reader, in the front line, in the shoes of the doctor at the bedside – to know that feeling of uncertainty and intrigue when confronted with a patient who has a problem that just might kill him. I try to show you the mind of the physician at work as she struggles to figure out exactly what is making the patient sick. To do this I have divided this book according to the steps we take in the evaluation of each and every patient we see. Each chapter focuses on one of the tools of our trade, how it's supposed to work, and how errors send us astray. As physicians become more open about what we do, we make it easier for patients to understand what they can do to more fully participate in their own care.

This book has its roots in a column I have written for the past six years for the *New York Times* magazine. The column has been my opportunity to share with general readers my personal collection of fascinating diagnostic histories. It's a collection I began (unwittingly) to assemble years ago, while my own medical career was still in its formative stages.

I came to medical school as a second career. The first I spent in television news, mostly covering medicine, mostly for CBS. I hadn't planned to go to medical school; it wasn't some long deferred dream. But one day, while filming with television correspondent Dr Bob Arnot, I watched him save an elderly woman's life. He was supposed to be shooting a stand-up on white-water rafting when he suddenly disappeared from the raft I was watching in the monitor. The cameraman and I searched the distance and saw him on the banks of the river, pulling an elderly woman onto the rocky shore. The cameraman refocused on this new image and I watched with fascination as Bob performed basic cardiopulmonary resuscitation (CPR) and brought the nearly drowned woman back to life.

I didn't quit television right then and there and head off to medical school, but it planted an idea and revealed a hidden dissatisfaction with my role in TV. Television reaches millions, but touches few. Medicine reaches fewer but has the potential to transform the lives of those it touches.

So I did two years of premed at Columbia University, then applied and was accepted at the Yale School of Medicine. I completed my residency training at Yale's Primary Care Internal Medicine programme and have stayed on here to care for patients and teach new generations of doctors.

When I started medical school I thought I would be most interested in pathophysiology – the science behind what goes wrong when we get sick.

And, in fact, I loved that subject and still do. But what captured my imagination were the stories doctors told about their remarkable diagnoses – mysterious symptoms that were puzzled out and solved. These were the stories I found myself telling my husband and friends at the dinner table.

Covering medicine for as many years as I had, I thought I understood how medicine worked. But these stories revealed a new aspect of medicine – one well known to doctors but rarely discussed outside those circles. In writing my columns and now this book, I try to share a face of medicine that is both exciting and important. Exciting because the process of unravelling the mystery of a patient's illness is a wonderful piece of detective work – complicated yet satisfying. Important because any one of us might someday be that patient. The more you know about the process, the better you will be prepared to assist and understand.

# PART ONE

## *Every Patient Tells a Story*

# CHAPTER ONE

## *The Facts, and What Lies Beyond*

The young woman was hunched over a large pink basin when Dr Amy Hsia, a resident in her first year of training, entered the patient's cubicle in the Emergency Department. The girl looked up at the doctor. Tears streamed down her face. 'I don't know if I can take this any longer,' twenty-two-year-old Maria Rogers sobbed. Since arriving at the emergency room early that morning, she'd already been given two medicines to stop the vomiting that had brought her there – medicines that clearly had not worked.

'I feel like I've spent most of the last nine months in a hospital or a doctor's office,' Maria told the doctor quietly. And now, here she was again, back in the hospital. She'd been perfectly healthy until just after last Christmas. She'd come home from college to see her family and hang out with her friends, and as she prepared to head back to school this strange queasiness had come over her. She couldn't eat. Any odour – especially food – made her feel as if she might vomit. But she didn't. Not at first.

The next day, on the drive back up to school, she'd suddenly broken into a cold sweat and had to pull over to vomit. And once she got started, it seemed like she would never stop. 'I don't know how I made it to school because it seems like I had to get out of the car to throw up every few minutes.'

Back at school she spent the first few days of the semester in bed. Once she was back in class her friends joked that she was just trying to get rid of the extra pounds from the holidays. But she felt fine and she wasn't going to worry about it.

Until it happened again. And again. And again.

The attacks were always the same. She'd get that queasy feeling for a few hours, and then the vomiting would start and wouldn't let up for days. There was never any fever or diarrhoea; no cramps or even any real pain. She tried everything she could find in the drugstore: Tums, Pepcid, Pepto-Bismol, Prilosec, Maalox. Nothing helped. Knowing that another attack could start at any moment, without warning, gnawed insistently in the back of her mind.

She went to the infirmary with each attack. The doctor there would get a pregnancy test and when it was negative, as it always was, he'd give her some intravenous fluids, a few doses of Compazine (a medicine to control nausea), and, after a day or two, send her back to the dorm. Halfway through the semester she withdrew from school and came home.

Maria went to see her regular doctor. He was stumped. So he sent her to a gastroenterologist, who ordered an upper endoscopy, a colonoscopy, a barium swallow, a CT scan of her abdomen, and another of her brain. She'd had her blood tested for liver disease, kidney disease, and a handful of strange inherited diseases she'd never heard of. Nothing was abnormal.

Another specialist thought these might be abdominal migraines. Migraine headaches are caused by abnormal blood flow to the brain. Less commonly, the same kind of abnormal blood flow to the gut can cause nausea and vomiting – a gastrointestinal equivalent of a migraine headache. That doctor gave Maria a medicine to prevent these abdominal 'headaches' and another one to take if an attack came anyway. When those didn't help, he tried another regimen. When that one failed, she didn't go back.

The weird thing was, she told Hsia, the only time she felt even close to normal during these attacks was when she was standing in a hot shower. Couldn't be a cold shower; even a warm shower didn't quite do it. But if she could stand under a stream of water that was as hot as she could tolerate, the vomiting would stop and the nausea would slowly recede. A couple of times she had come to the hospital only because she'd run out of hot water at home.

Recently, a friend suggested that maybe this was a food allergy, so she gave up just about everything but ginger ale and saltine crackers. And that seemed to work – for a while. But two days ago she'd woken up with that same bilious feeling. She'd been vomiting nonstop since yesterday.

Maria Rogers was a small woman, a little overweight with a mass of long brown hair now pinned back in a hair slide. Her olive skin was clear though pale. Her eyes were puffy from crying and fatigue. She looked sick, and was clearly distressed, Hsia thought, but not chronically ill.

How often did she get these bouts of nausea? she asked the girl. Maybe once a month, she told her. Are they linked to your periods? Hsia offered hopefully. The girl grimaced and shook her head. Are they more common just after you eat? Or when you're hungry? Or tired? Or stressed? No, no, no, and no. She had no other medical problems, took no medicines. She was a social smoker – a pack of cigarettes might last a week, sometimes more. She drank – mostly beer, mostly on the weekends when she went out with her friends.

Her mother had been an alcoholic and died several years earlier. After leaving college she had been living with her father and sister but a few months ago moved into a nearby apartment with some friends. She had no pets, had not travelled within the past year. Had never been exposed to any toxins as far as she knew. Hsia examined her quickly. The gurgling noises of the abdominal exam were quieter than normal and her belly was mildly tender, but both findings could simply be due to the vomiting. There was no sign of an inflamed gallbladder. No evidence of an enlarged liver or spleen. The rest of the exam was completely unremarkable. 'As I walked out that door,' Hsia explained to me, 'I knew I was missing something but I had no idea what it was. Or even what to look for.'

## *More Than Just the Facts*

Dr Hsia was a resident in Yale's Primary Care Internal Medicine residency training programme, where I now teach. She told me about Maria Rogers because she knew I collected interesting cases and sometimes wrote about them in my column in the *New York Times* magazine. In thinking about this case, Amy told me she knew from the start that if she was going to figure out what was causing this patient to suffer so, it wasn't going to be because she had greater knowledge – because Maria Rogers had already seen lots of experts. No, if she was going to figure it out, it would be because she'd find a clue that others had overlooked.

The patient's story is often the best place to find that clue. It is our oldest diagnostic tool. And, as it turns out, it is one of the most reliable as well. Indeed, the great majority of medical diagnoses – anywhere from 70 to 90 per cent – are made on the basis of the patient's story alone.

Although this is well established, far too often neither the doctor nor the patient seems to appreciate the importance of what the patient has to say in the making of a diagnosis. And yet this is crucial information. None of our high-tech tests has such a high batting average. Neither does the physical exam. Nor is there any other way to obtain this information. Talking to the patient more often than not provides the essential clues to making a diagnosis. Moreover, what we learn from this simple interview frequently plays an important role in the patient's health even after the diagnosis is made.

When you go to see a doctor, any doctor, there is a very good chance that she will ask you what brought you in that day. And most patients are prepared to answer that – they have a story to tell, one that they have already told to friends and family. But the odds are overwhelming that the patient won't have much of an opportunity to tell that story.

Doctors often see this first step in the diagnostic process as an interrogation – with Dr Joe Friday getting 'Just the facts, ma'am,' and the patient, a passive bystander to the ongoing crime, providing a faltering and somewhat limited eyewitness account of what happened. From this perspective, the patient's story is important only as a vehicle for the facts of the case.

Because of that 'facts only' attitude, doctors frequently interrupt patients before they get to tell their full story. In recordings of doctor-patient encounters, where both doctor and patient knew they were being taped, the doctor interrupted the patient in his initial description of his symptoms over 75 per cent of the time. And it didn't take too long either. In one study doctors listened for an average of sixteen seconds before breaking in – some interrupting the patient after only three seconds.

And once the story was interrupted, patients were unlikely to resume it. In these recorded encounters fewer than 2 per cent of the patients completed their story once the doctor broke in.

As a result, doctors and patients often have a very different understanding of the visit and the illness. Survey after survey has shown that when queried after an office visit, the doctor and patient often did not even agree on the purpose of the visit or the patient's problem. In one study, over half of the patients interviewed after seeing their doctor had symptoms that they were concerned about but did not have a chance to describe. In other studies doctor and patient disagreed about the chief complaint – the reason the patient came to see the doctor – between 25 to 50 per cent of the time. This is information that can come only from the patient and yet, time after time, doctors fail to obtain it. Dr George Balint, one of the earliest writers on this topic, cautioned: 'If you ask questions you will get answers, and nothing else.' What you won't get is the patient's story, and that story will often provide not only the whats, wheres, and whens extracted by an interrogation, but often the whys and hows as well.

Moreover, the interrogation model makes assumptions about the elicited symptoms and diseases. And while these assumptions might be true for most of the people with those symptoms, they may not be true for this particular individual. The great fictional detective Sherlock Holmes talks at length about the difference between the actions and thoughts of the individual when contrasted to the average. Holmes tells Watson that while you may be able to say with precision what the average man will do, 'you can never foretell what any one man will do.' The differences between the average and the individual may not be revealed if the doctor doesn't ask.

'It is much more important to know what kind of patient has the disease than what sort of disease the person has,' Osler instructed his trainees at the turn of the twentieth century. Even with all of our diagnostic technology and our far better understanding of the pathophysiology of disease, research suggests this remains true.

So getting a good history is a collaborative process. One doctor who writes frequently about these issues uses the metaphor of two writers collaborating on a manuscript, passing drafts of the story back and forth until both are satisfied. 'What the patient brings to the process is unique: the particular and private facts of his life and illness.' And what the physician brings is the knowledge and understanding that will help him order that story so that it makes sense both to the doctor – who uses it to make a diagnosis – and to the patient – who must then incorporate that subplot into the larger story of his life.

If getting a good history is so important to making an accurate diagnosis, why are we so bad at it? There are several reasons.

First, most researchers, doctors, and patients would agree that time pressures play an important role. A visit to a doctor's office lasts an average of twenty-two minutes. Although there is a sense that doctors are spending less time with their patients, that number has actually increased over the past twenty years. In 1989, the average doctor's appointment lasted only sixteen minutes. Despite this extra time, both doctors and patients frequently agree that their time together is still too short.

In response, doctors often depend on a few highly focused questions to extract the information they think will help them make a diagnosis quickly. Yet it is clear that this effort to reduce the time it takes to get a good history increases the risk of miscommunication and missed information. Like so many shortcuts, this information shortcut often ends up taking more time than those interviews in which patients are able to tell their stories in their own ways.

Studies suggest that getting a good history allows doctors to order fewer tests and make fewer referrals – without taking any more time. Indeed, some studies suggest that obtaining a good history can even reduce visit time. In addition, patient satisfaction is higher, adherence to therapy is higher, symptom resolution is faster, lawsuits are less frequent.

Lack of training may also contribute to the problem. Doctors spend two years in classrooms learning how to identify and categorise disease processes, matching symptoms to known disease entities, but until recently very few programmes offered any training on how to obtain that essential information. The assumption seemed to be that this did not need to be taught. And there may have been an unspoken expectation that our improved diagnostic technology would reduce our dependence on this kind of personal information. Studies have shown that neither assumption is true, and now most medical schools offer classes in doctor-patient communication. Moreover, since 2004, medical students are required to demonstrate proficiency in their history-taking skills in order to become licensed physicians. A new generation of physicians may not use these tools, but at least they have them.

Finally, many doctors are uncomfortable with the emotions that are sometimes associated with illness. When patients present their stories, they often look for cues from the doctor as to what type of information they should give. The interrogation format tells the patient that what's needed from them are the facts and only the facts. And yet illness is often much more than a series of symptoms. The experience of being sick is frequently interlaced with feelings and meanings that shape and colour a patient's experience and perception of a disease in ways that are unimaginable, and unanticipated, by the doctor. A family history of heart disease or cancer may lead a patient to minimise a symptom. I recently got a phone call from a friend, a man in his late fifties whose father had heart disease. My friend was having chest pains when he walked up a hill. He wondered if this was his childhood asthma returning. He was shocked when I suggested he see a cardiologist. He had two blocked arteries, which were opened with complete resolution of his pain. The same history might cause another to focus on a symptom well beyond its actual severity. I have a few patients who have had many stress tests because of their concerns over their chest pain. The fact that previous tests have not shown heart disease provides them with no comfort or reassurance. Financial concerns may likewise affect how patients tell their stories.

Worries about the social meaning of symptoms can complicate even a straightforward diagnosis. I learned this the hard way. A patient I saw when I was a resident came for a school physical. She was young and healthy. As I was finishing up and preparing to move on to the next patient, she suddenly asked me about a lesion on her buttocks. Could it be from doing sit-ups on the hard floor? she asked somewhat anxiously. I quickly looked at the lesion. It looked like a small blister, located in the cleft between the buttocks. Certainly, I reassured her, glancing at my watch. I noticed that she seemed worried about the lesion, but I didn't ask any further questions or do a more thorough exam because I was running behind schedule. Only when the lesion reappeared months later did she acknowledge that her boyfriend had had a breakout of genital herpes on a vacation they'd taken together and she hadn't insisted on his use of a condom. The reappearance of the lesions made herpes the likely culprit. I completely missed a straightforward diagnosis because I was too rushed to address her anxiety and she was too embarrassed to offer this other history. It happens all the time.

## *Everybody Lies*

Several years ago I got a call from a producer named Paul Attanasio. He had created a television show based in part on my column in the *New York Times* magazine and wanted to know if I would be interested in being a consultant for this new show. It was a drama, he told me, about an ornery doctor who was a brilliant diagnostician. I agreed to work on the show, thinking that it wasn't going to last long. The show, called *House M.D.,* quickly found an enthusiastic audience.

In this show, Dr Gregory House doesn't value patient history. Indeed, he frequently tells his trainees that they should not believe a patient's version of his illness and symptoms, because 'Everybody lies.' In the context of the show, there is a certain truth in that. Patients frequently lie to House and sometimes his staff – not because the patients are intrinsically deceitful but because of who House is. As portrayed (brilliantly by Hugh Laurie), House is far from the kindly and gentle doctor whose presence invites trust and confidences. Instead, he is narcissistic and arrogant, a drug addict, and something of a pedant. He is a darker, more bitter version of Conan Doyle's brooding detective Sherlock Holmes. House's demeanour tells patients that the feelings and meanings illness may have for them are not important and so they don't tell him about them. As a result, House often gets only part of their story.

The mystery is solved only when the rest of their story is revealed – either from evidence found when his staff break into the patient's home (a quirky twist on getting a thorough history) or when the patient is finally forced to reveal his hidden truths. House acknowledges the importance of a thorough patient history but concludes that the problem is the patient who lies rather than the doctor who fails to establish a relationship in which difficult, embarrassing, or distressing truths can be told.

Amy Hsia knew from the start that if she was going to figure out the cause of Maria Rogers's cyclic episodes of vomiting, it would be because of some key piece of history that she was able to get that others had overlooked. But sitting outside the patient's room that afternoon, she wasn't sure she'd found it. She went through the thick charts, reading the notes and test results collected by all of the other doctors involved in the same exercise in previous hospitalisations, trying to make it all make sense. Nothing leaped out at her. The sketchy description of the symptoms and history provided nothing she hadn't already found out from the patient herself.

Hsia considered the differential diagnosis once more. Nausea and vomiting have a very long list of causes: ulcers, gallstones, obstruction, infection. Hepatitis, pancreatitis, colitis, strokes, and heart attacks. None of them seemed to fit in a case of a young woman with multiple episodes of vomiting and lots and lots of tests showing no abnormalities. Maybe she wasn't going to be able to figure this patient out either. She ordered a new medication to relieve the nausea and then moved on to see her next patient.

The next morning, when Dr Hsia, her supervising resident, and the attending physician – the troika of the modern hospital medical patient care team – visited Ms Rogers, the girl's bed was empty. The sound of the shower told them where she'd gone. That caught the young resident's attention. When she had come by a couple of hours earlier to examine the girl, she'd been in the shower then too. She remembered that Rogers had told her that her nausea improved when she took a shower. What kind of nausea didn't get better with the traditional nausea medications – by now she'd been on most of them – but improved with a hot shower?

Hsia posed the question to the team. Neither had heard of such a syndrome. So, after the team had finished seeing all the patients they were caring for, Hsia hurried to find a computer. She went to Google and entered 'persistent nausea improved by hot showers.' She hit the enter key and less than a second later the screen was filled with references to a disease Hsia had never heard of: cannabinoid hyperemesis – persistent and excessive vomiting (hyperemesis) associated with chronic marijuana use (cannabinoid).

The disorder was first described in 1996, in a case report from an Australian medical journal. Dr J. H. Allen, a psychiatrist in Australia, described a patient admitted to his care with a diagnosis of psychogenic vomiting – vomiting due to psychological rather than physiological causes. Allen noticed that this patient's vomiting was associated with a bizarre behaviour – repetitive showering. He took a dozen showers each day. Allen also noticed that the symptoms improved over the course of his hospitalisation but recurred when the patient was sent home. The patient had a long history of chronic heavy marijuana use and Allen hypothesised that the vomiting might be triggered by the marijuana.

Over the next several years Allen noted similar patterns in other patients admitted with vomiting disorders, and in 2001 he published a paper reporting on ten patients with the disorder he named cannabinoid hyperemesis. Each patient in his series smoked marijuana daily; each had developed intermittent nausea and vomiting. All had used marijuana for years before they developed these episodic bouts of nausea and vomiting. And remarkably, nine of the ten patients reported that hot showers helped their symptoms when everything else failed. All symptoms resolved when these patients gave up marijuana. And then reappeared in three of the ten who resumed their cannabis use. Other case reports followed from around the world.

Could this be what was plaguing Hsia's patient? Did Maria Rogers even smoke marijuana? Hsia hurried back to the patient's room. She found the patient sitting in bed, a towel wrapped around her still wet hair. Yes, she did smoke marijuana frequently. Maybe not every day but most days. That clinched it – at least in Hsia's mind. The young doctor felt like cheering. She'd figured it out when even the experts had been stumped! This is really one of the great pleasures in medicine – to put the patient's story together in a way that reveals the diagnosis.

She excitedly explained to the patient what she'd found on the Internet – that there was a good chance that marijuana was causing her nausea. She got better in the hospital because she didn't use it when she was here. But when she got home and resumed her regular exposure to the drug, the nausea would once again be triggered. All she had to do was to give up smoking marijuana, Hsia concluded triumphantly, and her symptoms would be cured forever.

This story, which seemed so logical and reasonable from Hsia's perspective, did not make the same kind of sense to the woman who was living it every day. Rogers's response was immediate and emphatic and – to Hsia – shocking. 'That is total bullshit. I don't buy it,' the patient snapped angrily. She knew many people who used marijuana a lot more than she did and they didn't get sick like this. How could Hsia explain that? Huh? Besides, wasn't marijuana supposed to help people who were sick from chemotherapy? Why would it decrease nausea in that case and cause nausea in her? she demanded. Where was her proof? Where was her evidence?

Hsia was taken aback by the patient's anger. She thought the young woman would be thrilled by the news that simply stopping the marijuana use would cure her of this devastating illness. Why was she so angry?

Later that morning, Hsia told the attending and resident what she'd found and how angry the patient had become when she told her about this diagnosis. It made sense to the other doctors caring for the patient. The marijuana use, the cyclic nature of the symptoms, and the restorative powers of the hot shower made it seem like a slam dunk. But how were they going to convince the patient?

They never got the chance. Maria Rogers left the hospital the following day. When contacted several weeks later, Rogers reported that the nausea had recurred. Yes, she had resumed her usual practice of smoking marijuana most days because she still didn't believe there was a link. She had arranged for an evaluation by a gastroenterologist at Yale. When I spoke with Ms Rogers afterward, she told me that the doctors there had ordered many of the same tests her previous doctors had done. Not surprisingly, the results were no different. From Maria's perspective, what she had was still a mystery.

In medicine, the patient tells the story of his illness to the doctor, who reshapes the elements of that story into a medical form, into the language of medicine. The doctor will usually add to the story, incorporating bits of information gleaned through questions, from the examination of the body, from the tests that have been performed – and the result should be a story that makes sense – where all pieces ultimately add up to a single, unifying diagnosis.

But the story of the illness can't stop there. Once the diagnosis is made, the doctor has to once again reshape the story she has created – the story that helped her make the diagnosis – into a story she can then give back to the patient. She has to translate the story back into the language and the context of the patient's life so that he can understand what has happened to him and then incorporate it into the larger story of his life. Only when a patient understands the disease, its causes, its treatment, its meaning, can he be expected to do what is needed to get well.

Studies have repeatedly shown that the greater the patient's understanding of his illness and treatment, the more likely it is that he will be able to carry out his part in the treatment. Much of this research has been done in patients who have been diagnosed with diabetes. Patients who understand their illness are far more likely to follow a doctor's advice about how to change their diet and how to take their medications than those who do not.

It's understandable. Taking medications on a regular basis isn't easy. It requires dedication on the part of the patient. Motivation. A desire to incorporate this inconvenient addition into a life that is already complicated. Greater understanding by the patient has been shown to dramatically improve adherence. This is where getting a good history – one that provides you with some insight into the patient and his feelings about his illness, his life, his treatment – can really pay off.

To go back to the story of Maria Rogers, Hsia told me how surprised she was when the patient didn't accept her explanation of her illness. That marijuana was linked to the nausea and vomiting seemed obvious to Dr Hsia. It was not obvious to Ms Rogers. Perhaps there was no way for Hsia to explain this to her that would have been acceptable. The story Hsia told to this patient was the doctor's story – the observations and research that allowed Hsia to make the diagnosis. What she didn't do was create the patient's version of the story – one that would make sense in the larger context of her life.

And then the patient left the hospital and with her their chance to figure out how to help her understand her illness. Dr Hsia tried to stay in touch with Maria after she left the hospital, but after several months the cell phone number she gave was disconnected and a letter was returned. And so, having rejected one diagnosis and the treatment option it suggested, Maria Rogers still suffers from a malady for which she has no name and no cure.

## *Stories That Heal*

One of the most important and powerful tools a doctor has lies in her ability to give a patient's story back to the patient, in a form that will allow him to understand what his illness is and what it means. Done successfully, this gift helps the patient incorporate that knowledge into the larger story of his life. Through understanding, the patient can regain some control over his affliction. If he cannot control the disease, he can at least have some control over this response to the disease. A story that can help a patient make sense of even a devastating illness is a story that can heal.

The primary work of a doctor is to treat pain and relieve suffering. We often speak of these two entities as if they were the same thing. Eric Cassell, a physician who writes frequently about the moral dimensions of medicine, argues, in a now classic paper, that pain and suffering are very different. Pain, according to Cassell, is an affliction of the body. Suffering is an affliction of the self. Suffering, writes Cassell, is a specific state of distress that occurs when the intactness or integrity of the person is threatened or disrupted. Thus, there are events in a life that can cause tremendous pain, and yet cause no suffering. Childbirth is perhaps the most obvious. Women often experience pain in labour but are rarely said to be suffering.

And those who are suffering may have no pain at all. A diagnosis of terminal cancer, even in the absence of pain, may cause terrible suffering. The fears of death and uncontrollable loss of autonomy and self combined with the fear of a pain that is overwhelming can cause suffering well before the symptoms begin. There are no drugs to treat suffering. But, says Cassell, giving meaning to an illness through the creation of a story is one way in which physicians can relieve suffering.

In the case of Maria Rogers, Dr Hsia was able to gather the data necessary to make a diagnosis. She knew the disease the patient had. And yet she didn't know enough about the person who had the disease. The story she gave back to the patient was a reasonable one and a rational one, but it was not one the patient could accept. And when confronted with the vehement rejection of that story and the raw emotion displayed, Hsia retreated. Before she was able to regroup and try again, the patient left her care. Rogers rejected Hsia's story, rejected her diagnosis, and, when last I spoke with her, continued to search on her own for an end to her pain and suffering.

And yet the right story has nearly miraculous powers of healing. A couple of years ago I got an e-mail from a patient whose remarkable recovery highlighted the difference between pain and suffering and the healing power of the story. Randy Whittier is a twenty-seven-year-old computer programmer who was in perfect health and planning to get married when suddenly he began to forget everything. It started one weekend when he and his fiancée travelled to her hometown to begin making the final arrangements for their wedding the following spring. He had difficulty concentrating and was frequently confused about where they were going and whom they were talking with. He chalked it up to fatigue – he hadn't been sleeping well for some time – and didn't say anything to his fiancée. But on Monday morning, when he went back to work, he realised he was in trouble and sent an instant message to his fiancée, Leslie.

Leslie saw the flashing icon on her computer announcing that an instant message had arrived. She clicked on it eagerly.

'Something's wrong,' the message read.

'What do you mean?' she shot back.

'My memory is all f'ed up. I can't remember anything,' he wrote. Then added: 'Like I can't tell you what we did this weekend.'

Leslie's heart began to race. Her fiancé had seemed distracted lately. She thought maybe he was just tired. But he'd been strangely quiet on their trip to New York this weekend. He had been excited when they set up the trip, and she'd worried that he was getting cold feet.

'When is our wedding date?' she quizzed him. If he could remember anything, he'd be able to remember that. Planning this wedding had dominated their life for the past several months. 'Can you tell me that?'

'No.'

'Call the doctor. Do it now. Tell them this is an emergency.'

Over the next half hour, Randy put in three calls to his doctor's office, but each time he had forgotten what they told him by the time he messaged his fiancée. Separated by miles of interstate and several suburbs, Leslie was frantic. Finally, at her insistence, Randy, now terrified, asked a friend to take him to the closest hospital.

A few hours later, her cell phone rang. At last. He was being discharged, he told her. The emergency room doctor thought his memory problems were caused by Ambien, the sleeping pill he was taking. The doctor said the symptoms would probably improve if he stopped taking the medication.

Leslie didn't buy that for a second. 'Don't go anywhere,' she instructed him. 'I'll pick you up. I'm going to take you to your doctor.' A half hour later she found Randy wandering down the street in front of the hospital, uncertain about why he was there and even what her name was. She hustled him into the car and drove to his doctor's office. From there they were sent to Brigham and Women's Hospital in Boston.

Late that night, the on-call resident phoned Dr William Abend at home to discuss the newest admission. Abend, a sixty-one-year-old neurologist, scrolled through the patient's electronic medical record as the resident described the case. The patient, who had no history of any previous illnesses, had come in complaining of insomnia and severe memory loss. Psych had seen him – he wasn't crazy. His physical exam was normal except he didn't know the date and he couldn't recall the events of the week or even that day. The ER had ordered an MRI of his brain but it hadn't been done yet.

The patient needed a spinal tap, Abend instructed, to make sure this wasn't an infection, and an EEG, an electroencephalogram, to see if he was having seizures. Both could affect memory. He'd see the patient first thing the next morning.

Randy was alert and anxious when Abend came to see him. Tall and slender with earnest blue eyes, the young patient seemed embarrassed by all that he couldn't remember. His fiancée had gone to get some rest, and so his mother provided the missing details. He'd first complained about some memory problems a couple of months earlier. The past weekend everything got much, much worse. He couldn't remember anything from the past few days. He couldn't even remember he was in the hospital. Overnight, he repeatedly pulled out his IV.

On exam, Abend found nothing out of the ordinary save the remarkable degree of short-term memory loss. When Abend asked the patient to remember three words – automobile, tank, and jealous – the patient could repeat them but thirty seconds later he could not recall even one. 'It wasn't like – where did I put my car keys?' Abend told me. 'He really couldn't remember anything.' The neurologist knew he had to determine what was going on quickly, before further damage was done.

Abend checked the results of the spinal tap – there were no signs of infection. Then he headed over to radiology to review the MRI. There was no evidence of a tumour, stroke, or bleeding. What the MRI revealed were areas that appeared bright white in the normally uniform grey of the temporal lobe on both sides of the brain.

There are only a few diseases that would cause this kind of injury. Viral encephalitis – an infection of the brain that is often caused by herpes simplex – was certainly the most common. Autoimmune diseases like lupus could also cause these kinds of abnormalities. In lupus, the body's natural defenses mistakenly attack its own cells as if they were foreign invaders. Finally, certain cancers can do this too – it's usually lung cancer, usually in older smokers.

The young man's symptoms had been coming on gradually over two months. Abend thought that made an infection like herpes less likely. The patient had already been started on acyclovir – the drug usually used to treat herpes encephalitis – since the disease can be deadly when it infects the brain. Although Abend thought it unlikely, they would need to do additional tests of the spinal fluid to make sure there was no evidence of this dangerous viral infection.

Lupus seemed even more unlikely to Abend. It is a chronic disease that can attack virtually any organ in the body and is generally characterised by joint pains and rashes. The patient had none of these symptoms. Still, perhaps this was the first sign of this complex disease. It would be unusual, but so was the young man's extensive memory loss.

Although cancer was an uncommon cause of this kind of injury, it seemed to Abend the most credible in this patient. Even nonsmokers can get lung cancer. And other cancers can cause the same type of brain injury. Moreover, if these symptoms were caused by a cancer, there was a good chance that they would resolve once the cancer was treated. He ordered a CT of the chest, abdomen, and pelvis. Ordering all of these scans communicates uncertainty about what you are looking for and where it might be located, but Abend felt strongly that they didn't have time to be wrong.

Results from the tests trickled in over the next few days. He wasn't having seizures. It wasn't a virus. He didn't have lupus. But by the time those test results arrived they already had an answer. The CT of Randy's chest had shown a large mass – not in his lungs, but in the space between them, the area called the mediastinum. A biopsy revealed the final diagnosis – Hodgkin's lymphoma, a cancer that attacks the immune system. He had what is called a paraneoplastic syndrome, a rare complication in which antibodies to his cancer attacked the healthy cells in his brain.

Randy had surgery to reduce the size of the mass and then started chemotherapy. And slowly, remarkably, his memory began to improve. But the trip to New York remains vague, and his only memory of his weeklong hospital stay is his nurse telling him he was going home.

His fiancée remembers the day she realised he was getting better. It was several weeks after leaving the hospital. She reminded him that he wanted to get a haircut. He told her that he tried to go the day before but the line at the barbershop was too long.

She almost cried. 'At that moment,' she told me, 'I finally knew that the man I loved was still in there and that he was coming back.'

When I called Randy after receiving his e-mail, he still couldn't remember much of his ordeal, but he understood the illness and the prognosis. One doctor stood out from the crowd of physicians caring for him. Marc Wein was a medical student at Brigham, and he had become fascinated by Randy and his illness. He read voraciously about the disease, tracked down case reports of other patients with a similar manifestation of cancer, and came back again and again to explain it all to Randy and Leslie. Together Marc and Randy created the story of this remarkable diagnosis that made sense to both of them. And that made all the difference.

Randy tells me he was never in pain but he hated the way he became a clean slate every five minutes. He hated the worried looks he saw on the faces of those he loved. He hated the loss of a sense of who he even was.

He embraced the story that Wein put together for him. Leslie had to remind him frequently of the particulars of that story, but he remembered that he had a cancer and that curing that cancer would restore him to himself. He welcomed the surgery and never minded the pain from the incision down his chest. He even looked forward to chemotherapy. Watching the intravenous needle pierce his skin, he remembered it meant he was one step closer to getting better. I spoke with Randy several times as he faced his ordeal. His optimism never flagged. He is now disease free and his life has moved on. He returned to work five months after that strange weekend and got married the next year.

Randy's body may have been cured by the chemotherapy, but his mind was healed by a story.

# CHAPTER TWO

## *The Stories They Tell*

At a recent conference of the American College of Physicians in Philadelphia, a friend who knew of my interest in diagnosis encouraged me to attend one lecture in particular. The title stood out from all the Updates on Cardiology and Innovations in Nephrology, Haematology, or Urology. This talk was called simply 'Stump the Professor.'

When I arrived at the designated ballroom I was amazed – the place was packed with hundreds of doctors. As I picked past feet and knees to claim a rare open seat, I looked at the casually dressed, mostly middle-aged audience. There was a sense of giddy anticipation in the air, reminiscent of a college-age trek to a distant concert arena.

Finally, a tall and slender woman with a volleyball of grey curls and a broad smile strode onto the stage, nodding and smiling at her devotees. The audience exploded with applause.

This was Dr Faith Fitzgerald, a flesh-and-blood version of TV's Dr House. She is the doyenne of the diagnostic dilemma. This auditorium of hundreds of doctors had come to see her take on a series of challenging cases – patients whose stories had been submitted by medical students from around the US and handpicked for this presentation because of their difficulty and complexity. The patient's story and medical course would be presented to Fitzgerald, a bit at a time, and her job would be to figure out the diagnosis by the end. Throughout the presentation she would take the audience through her thought process, acting the modern Sherlock Holmes to her own crowd of Dr Watsons. It was another mark of our time: diagnosis was now a form of entertainment.

After what appeared to be a completely unnecessary introduction to this crowd, Fitzgerald set her glasses halfway down her long, aquiline nose, and greeted the adoring fans. Like all good speakers, she started off with a joke – a doctor joke: 'Before we get started, and just for the record,' Fitzgerald growled in her tobacco-raspy voice, 'I'd like to mention – endocarditis, tuberculosis, Wegener's granulomatosis, Kawasaki's aortitis, Jakob-Creutzfeldt dementia, and eosinophilic gastritis.' She rushed through this list of arcane diseases and ended with a laugh. 'I don't know any of the cases I'm about to hear but there's a darn good chance I've mentioned at least one case diagnosis in that list. Just so you know that I did say them.'

The crowd laughed appreciatively. In this forum, even if you don't ultimately figure out the case, you get credit for having the final diagnosis among the diseases you considered on the way. Fitzgerald was acknowledging that the cases she would be likely to confront that day would not be the same as those doctors routinely see in daily practice. Instead they would be the 'fascinomas,' the intriguing cases physicians share at the watercooler, the nurses' station, or in hospital stairwells.

Javed Nasir, a twenty-something graduating medical student from the Uniformed Services University Medical School, walked onto the stage. He would present the first case – a patient he cared for in his third year. 'Good morning, Dr Fitzgerald.' His voice wavered slightly. He began, with what is called (by tradition) the chief complaint. ' "My wife is not acting right." ' The young man looked out at the large crowd uncertainly and then continued. 'This is a seventy-three-year-old woman with a three-month history of progressive confusion, brought to the hospital by her husband.' He then detailed the patient's symptoms in the conventional medical format.

Over the next ninety minutes these doctors watched and occasionally helped Fitzgerald work her way through Nasir's and two other patient stories, revealing through each the internal machinery of making a diagnosis.

She had never met any of these patients, had never examined them. Instead Fitzgerald made her diagnosis using a doctored-up version (quite literally) of the patient's story. That story contained only the barest bones of the original patient's story, stripped of all that is unique, personal, and specific, reshaped by the doctor and augmented by the physical exam findings and test results from the investigation. All this was presented in a highly structured and familiar format.

Although this is done as a kind of entertainment, a kind of brain-teaser for the audience full of doctors, it's a simulacrum of what doctors do at the bedside. The kind of stripped-down and highly structured story on which this exercise depends is one of the most important tools doctors have for translating the abstract knowledge of the body – gleaned from cadavers, test tubes, and books – into a diagnosis of the patient before them. It is a familiar exercise to doctors because we are the authors of these stories for our own patients and audience for other doctors seeking help with theirs.

Nasir continued with his patient's story, explaining that she had been in her usual state of health until a few months earlier, when she became increasingly forgetful. First, she began to have trouble finding the right words when she spoke. Her husband got really scared when she started to get lost driving even in her own neighbourhood. At the time of her admission she was having difficulty with the most basic daily activities; she could no longer cook or even dress herself without his help. She was unwilling even to leave the house without him.

Fitzgerald is an internist, and a dean of medicine and humanities at the University of California at Davis. As the medical student told this patient's story of rapidly worsening confusion she paced up and down the stage. Her long black coat flapped behind her, revealing slim black trousers and black turtleneck – her usual attire.

An old hand at this format, she was clearly enjoying the challenge and the crowd – a mixture of old hands and novice trainees. Fitzgerald has been a regular feature at conferences like this one for more than a decade.

'On physical exam, the patient is a thin, frail woman who appeared timid and fearful,' Nasir continued.

'Timid and fearful?' Fitzgerald asked. (In the movie version, she might puff on her calabash about now.) 'Hmmm. That could be part of her confusion or could be her personality. Did you get a sense of what she was like before all this?' The student shook his head. 'Well, it would certainly be hard to feel confident in a world that you suddenly don't understand.'

The rest of the physical exam was unremarkable, the medical student told her.

Fitzgerald stopped pacing. 'By that I guess you mean that it was normal?' she asked.

Nasir nodded. 'Even the neurological exam – completely normal?' Again he nodded. Fitzgerald was silent as she considered the story so far.

'Would you like to order some tests?' the student prompted. In this structured performance the doctor can ask for any test and if the patient had the test that data will be shared.

'Sure.' She quickly called out tests she'd like to order and the results were provided. A spinal tap was normal, there was no elevated white blood cell count, her liver and kidneys were working fine.

'So basically what you're telling me is that we have here a woman with a rapidly progressive dementia but a completely normal physical exam otherwise and no sign of infection or laboratory abnormalities?' Fitzgerald asked. She then turned to the audience. 'I am not at all offended if people shout out the answer at any time,' she called out to the audience. 'Anyone? Well, at least it's not obvious to anyone else out there either.'

It certainly wasn't obvious to me. As Fitzgerald considered the data available on the patient, she started to describe how she was thinking about what she'd heard. 'At this point I like to develop some kind of structure on which to hang my ideas. To help me put together a thorough differential diagnosis, I often just start with the different areas of medicine. So, could this be some kind of congenital disease that causes dementia – like early Alzheimer's? Maybe. Or could this be infectious? Did she have a life of adventure that would put her at risk for some colourful, sexually transmitted diseases like syphilis or HIV?'

As she reviewed her thinking, she developed a list of possible causes of these symptoms. Voices called out from the audience offering additional diseases to add to the differential. 'Parkinson's dementia' a man called from the end of my row. 'Jakob-Creutzfeldt' (mad cow disease) offered a woman in front.

'Get a head CT,' called out still another voice.

'Hmmm – a head CT.' Fitzgerald considered the suggestion. 'This lady has no neurological findings – right?' She turns to Nasir, who again nods his confirmation. 'No weakness, no seizures, no tremor – nothing except confusion. Given that, I don't think a CT scan will show me much. In my hospital it's almost impossible for a patient with mental status changes to come through the ER without getting a head CT. And yet the odds are that hers will be normal, so . . .' She paused thoughtfully. 'I say we skip it.'

Once the case had been presented completely, it was time for Fitzgerald to make her diagnosis. She went through her differential. 'Well, common things being common, this would most likely be multi-infarct dementia or maybe Alzheimer's. But this is stump-the-professor time and so it's never the common thing. Hmmmm.' She turned to the audience. 'Can I talk to a really old doctor?' Chuckles from the audience were followed by a few more suggestions.

'Any other ideas?' Fitzgerald conceded. 'Okay, I give up. Let's hear it.'

'Maybe you should have gotten the head CT after all,' quipped the medical student, pleased that he actually stumped the professor. He projected the final slide onto the large screens at the front of the room. An image from a CT scan of the head revealed a huge, white, irregularly shaped circle bulging into and distorting the familiar spaghetti swirls of the brain. It was a brain tumour.

'Damn. It's big too,' conceded Fitzgerald, shaking her head. 'It's really amazing that it didn't announce itself more clearly. Oh well, you can't win them all, now can you?' she said, facing the audience with a roguish smile. The audience applauded enthusiastically.

I turned to the young woman sitting next to me, still clapping. 'Aren't you disappointed that she got it wrong?' I asked. She shook her head. 'No way. This is about the process – hearing the story and putting it all together.

I started off wanting to be a surgeon, but I realised that it was internal medicine that would keep me on my toes intellectually.'

The man sitting next to her leaned over and added, 'I didn't come here for the answer. I come to see the thinking.'

Getting the right diagnosis is, of course, what you always want – and will usually get on TV and in the movies. But doctors are hungry to hear how others think a case through. Translating the big, various, complicated, contradictory story of the human being who is sick into the spare, stripped-down, skeletal language of the patient in the bed, and then making that narrative reveal its conclusion – that is the essence of diagnosis. Like a great Hitchcock film, the revelation at the end is not nearly as interesting as the path that gets us there. So despite her wrong answer, it was exciting to watch Fitzgerald work her way through this complicated case. And, in the other two cases presented that afternoon, she was right. I caught up with Fitzgerald later that day. 'Oh, I'm wrong a lot, but my audience seems to forgive me.' Fitzgerald laughed, then added, 'It's a form of entertainment. A lot of the appeal of internal medicine is Sherlockian – solving the case from the clues. We are detectives; we revel in the process of figuring it all out. It's what doctors most love to do.'

The kind of story Javed Nasir told to Fitzgerald is at the very heart of that Sherlockian process. It is one of the fundamental tools of diagnosis. Doctors build a story about the patient in order to make a diagnosis. It is a story based on the patient's story but it is freed of most of the particular details of the individual, and structured to allow the recognisable pattern of the illness to be seen. In the last chapter I looked at the process of getting the story from the patient and the final task of giving it back to the patient. Here I want to look at just what it is that doctors do with that story to make it yield the diagnosis.

Done well, the doctor's version of the story often holds the key to recognising the pattern of an illness, leading to a diagnosis. Much of the education doctors get in their four years of medical school and subsequent years of apprenticeship training is focused on teaching this skill of identifying and shaping those aspects of a patient's life and symptoms, exams and investigations that contribute to the creation of a version of the patient's story that makes a diagnosis possible. Indeed, the ability to create this spare and impersonal version of the patient's story is *the* essential skill in diagnosis.

It's also one of the aspects of medicine that can seem most dehumanising. It's how the elegant retired schoolteacher who mesmerised three generations of her students with stories of the Roman Empire as she inspired them to master noun declensions in Latin is quickly reduced, in diagnosis-speak, to the seventy-three-year-old woman with rapidly progressive dementia in room 703.

How doctors apply general medical knowledge to the particular patient has been an area of intense interest and research for decades. Current thinking focuses on stories as the key. The basic sciences of anatomy, physiology, biology, and chemistry are linked to a patient at the bedside through very specific stories that doctors learn and eventually create. These stories, what researchers now call illness scripts, contain key characteristics of a disease to form an iconic version, an idealised model of that particular disease. For any individual disease, the illness script will be a loosely organised aggregate of information about the typical patient, about the usual symptoms and exam findings – with an emphasis on those that are unique or unusual – as well as information about the pathology and biology of the disease itself. It is the story that every doctor puts together for herself with the knowledge she gains from books and patients. The more experience a doctor has with any of these illnesses, the richer and more detailed the illness script she has of the disease becomes.

Development of a large library of these illness scripts has been the goal of medical training since long before it was described this way. When I was a student and then a resident in the 1990s, you'd hear older doctors tell you that the only bed you couldn't learn from was your own. That's why residency programmes exist. Seeing more patients helps you learn more medicine and become a better doctor.

One of the ways doctors are taught to think about disease, one of the ways that these illness scripts get structured, is through the use of what are known as clinical pearls – observations and aphorisms containing nuggets of information about patients and likely diagnoses. This is a teaching technique that dates back to the days of Hippocrates, who published several volumes simply titled *Aphorisms*. Modern medical students are drilled on the five Fs of gallbladder disease – female, fat, forty, fertile, and fair – the characteristics of the most typical patient. They are pumped on Charcot's triad – fever, jaundice, and right upper quadrant pain (the diagnostic trio of a gallbladder infection that is spreading to the liver).

Clinical pearls are often cleverly worded to make it easier for students to remember them. When taking care of a patient who came in with a paralysed arm and a facial droop I was told: a stroke is only a stroke after 50 of D50 – a reminder that low blood sugar (which can be treated with 50 mg of 50 per cent dextrose, or D50) can cause symptoms that imitate those of a stroke. When I was seeing a patient in the ER brought in after being found in a snowbank, a patient who had no detectable heart rate or blood pressure, I was told: a man isn't dead until he's warm and dead. That is, in conditions of extreme hypothermia (low body temperature), vital signs may be undetectable until the body temperature is brought up to a near normal range. And in fact this patient recovered fully. These pearls are little snippets of the illness script, snippets that help doctors connect a patient to a diagnosis.

Doctors create stories about patients that are organised like these illness scripts. Using the barest most generalised recounting of the patient's characteristics, his symptoms, his exam and test results, the doctor tries to match that story to an illness script in order to make a diagnosis, or at least build a differential. A well-constructed story might even help a doctor who has never seen a patient to come up with the right diagnosis.

Tamara Reardon is alive today because a doctor – not *her* doctor – was able to make a diagnosis based on a one-line description of her illness. Tamara was forty-four years old, a mother of four, and healthy until one day in early spring when she woke up with a sore throat and a fever. She took some Advil, got her children off to school, and went back to bed. She was still there when the kids got home that afternoon. She roused herself enough to get them started on their homework, then returned once more to bed. Her entire body ached; she alternated between shuddering chills even under a half dozen blankets and waves of heat marked by drenching sweats. Her husband made dinner that night but she couldn't eat. The next day she could barely drag herself out of bed to see her doctor. She still had a fever, her throat was on fire, and she had a new symptom: her jaw hurt, mostly on the right, so that talking and eating were excruciating. When the doctor had her open her mouth so he could look at her throat, it hurt so much she cried.

Tonsillitis was his diagnosis. Probably strep throat. An outbreak had roared through her household a few weeks before, so the doctor didn't even send a culture. He simply sent her home with a prescription for an antibiotic called Biaxin. After a couple of days of antibiotics Tamara began to feel better. The fever came down and her throat was less painful, but now she noticed a lump in her neck that had her worried. She went back to her doctor. He looked down her throat. It was much easier this time – her jaw was no longer painful. Her tonsils looked fine – the fiery red colour was gone and they no longer looked swollen. But across the back of her throat the doctor saw patches of white that hadn't been there before. And her neck was swollen and tender on the right. The doctor thought the swelling was probably just a lymph node still inflamed from her recent infection, but he was a little puzzled about the white patches. He gave Tamara a week's worth of prednisone – a steroid – to reduce the inflammation since it was bothering her. And he made an appointment for her to see an Ear, Nose, and Throat doctor about those white patches.

The steroids reduced the swelling in her neck almost immediately. And the fatigue and achy feeling she'd had since she'd first become sick started to ease up. Whatever she'd had, it was gone now.

The day after she'd taken her last dose of prednisone, she woke up with a fever. And the swelling on her neck was back – and even worse than it had been before she'd taken the steroids. She could hardly open her mouth.

She could not move her neck. She had an appointment with the ENT the next day, but Tamara felt too sick to wait. Her husband drove her to the emergency room and after waiting a couple of hours she was given some Darvocet (a painkiller) and advised to see her ENT the next day.

She did, but he wasn't certain what was going on either. She had a fever and her neck was swollen and red on the right. It seemed too extensive to simply be her lymph nodes. He worried she might have an abscess hidden in her tonsils. The white patches her doctor had been worried about were gone. He looked into her throat using a tiny camera embedded at the end of a slender tube. He couldn't find any evidence of an abscess, so he gave her a few more days of steroids and another round of antibiotics. And he got a CT of her neck.

That night the ENT went to a meeting of his local medical society. He ran into an old friend, Dr Michael Simms, a specialist in infectious disease. As they made their way to their seats, the ENT thought of this baffling case. 'Hey Mike, let me run something by you. I've got a forty-four-year-old woman with a history of tonsillitis who now has fever, jaw pain, and swelling on the right side of her neck. I got a CT scan and there's no abscess, just a clot in the jugular vein. Do you know what this is?' Simms looked at his friend. He ticked through the facts the ENT had related: 'She had a recent case of tonsillitis, and now has fever and pain in the right side of her neck and a clot in her jugular vein?' The ENT nodded. 'I think she has Lemierre's disease,' Simms told him, instantly.

Dr André Lemierre, a physician in Paris, first described this disease in 1936. It's rare, and seen most often in adolescents and young adults. Lemierre wrote up several cases of this illness, which begins with a fever and tonsillitis and progresses to a painful and often swollen neck as the infection moves into the jugular vein. Once there, the bacteria induce the formation of blood clots, which then shower the rest of the body with tiny bits of infected tissue.

Before the discovery of penicillin the disease was usually fatal. The widespread use of penicillin to treat all severe sore throats during the 1960s and 1970s virtually wiped out the disease. But over the past twenty years,

Lemierre's has staged something of a comeback – an unintended consequence of a more cautious use of antibiotics and the development of new drugs – like Biaxin, which is what Tamara was given – that are easier to take but far less effective than penicillin against this potentially deadly infection.

Simms saw Tamara the next day. Since starting the medications she felt much better – hardly sick at all – so she was surprised when Simms recommended that she go to the hospital that very day. She went, and just in time. The infection had already moved into her lungs. She had a rocky course, and ended up spending nearly two months in the hospital – but she survived.

With only a couple of sentences, and a handful of facts about the case, Michael Simms was able to diagnose this woman he had never seen, a patient whose diagnosis had already been missed by two primary care doctors and a doctor specialising in diseases of the head and neck. That is the power of these little stories.

Clearly, knowledge is an important part of this. Simms was able to make this diagnosis because he knew this disease. It's rare, so it's likely that the patient's primary care doctor and the ER doctor had never heard of it. But the ENT knew about this disease. When Simms mentioned Lemierre's, he'd recognised it. But somehow he hadn't been able to connect the knowledge of the disease with its classic clinical presentation. Somehow he hadn't created a story or illness script for this entity. Maybe he'd never seen it before either. I doubt he'll miss it again.

Doctors are constantly adding to the number and richness of the illness scripts in their heads. Every patient contributes. Lectures can too. Most speakers start off with a classic patient story before presenting their research on a disease or topic. Medical journals often present difficult cases in their pages. Like those presented to Fitzgerald, these cases teach doctors about a particular disease, and about the construction of the story that can help the doctor link the patient to the diagnosis.

These stripped-down stories, while useful to the diagnostic process, bear little resemblance to the stories a patient tells the doctor. Doctors strip away the personal and specific to make their version of the story and in doing so sometimes forget that the reason we do this is to help the person in the bed. That person is more than their disease, but sometimes that seems to get forgotten. When doctors confuse the story they have created about the patient's disease with the patient himself, this contributes to a sense that medicine is cold and unfeeling and indifferent to the suffering of patients – the opposite of what medicine should be.

Dr Nancy Angoff is the dean of students at Yale Medical School. She watches over the one hundred students of each class as they wend their way from student to doctor. She's concerned that medical education spends too much time on focusing the students' attention on the disease and not enough time on the patient. She cringes when she overhears a student refer to a patient by his disease and location, or when the discussion of a cool diagnosis overlooks the potentially tragic consequences for the person with the diagnosis. She worries that the doctors they will become will forget how to talk to the patient, to listen to the patient, to feel for the patient. For years she worried that in the excitement of mastering the language and culture of medicine they might lose the empathy that brought them to medical school in the first place.

When Angoff became the dean of students, she decided to see if she could do something to prevent that transformation. And she wanted to do it right from the start, right from the very first day of school. 'Students come here and they are very excited about medicine. They want to help the sick patient, and medicine is the tool that makes that possible. That's why they are here. But medical schools don't teach you about the patient, they teach you about the disease. I wanted to emphasise the patient right from the very first day.'

As part of that effort, Angoff has shaped that first day at Yale Medical School to try to 'vaccinate' the students against the focus on the disease and the depersonalisation of the patient that is part and parcel of current medical education. To do this, she focuses on the difference between the patient's story and the story the doctors create from it.

So on a warm September morning, I returned to the classroom in which I had spent most of my first two years as a medical student to see what a new generation of med students is taught about the stories we hear and those we tell as doctors.

As Angoff, a small and slender woman in her mid-fifties, stepped onto the stage, the nervous chatter of these brand-new students quickly died. She said a few words of welcome and then outlined the events of the morning. We would hear two versions of a patient's story, first as the patient told it and then as it might have been written up by a doctor caring for the patient in the hospital.

The stories were to be performed by Dr Alita Anderson. Anderson is a young black woman in her early thirties. A Yale Medical School graduate (class of 2000), Anderson spent a year interviewing patients about their experiences in the health care setting. All of the patients she interviewed were African American, most were poor; many were poorly educated as well. All had multiple encounters with a medical system that was only sometimes responsive to their needs. She now travels around the US performing the stories she collected from this often unheard population.

Anderson gave Angoff a hug and then walked slowly across the stage. She began to sing a slow sad song in a husky alto. I couldn't quite understand the words and I didn't recognise the song, but it sounded like some kind of spiritual.

Anderson settled in a lonely chair on the stage and finished the song. She sat quietly for a moment and then said in a rumbly southern voice, 'In June 1967, I went to Vietnam. I was a member of the First Infantry Division. My first evening there, they sent me out on an ambush.' She didn't have any props, nor a costume, but through her voice and expressions she became this middle-aged black man who never recovered from the battlefields and bars of his year in the Vietnam War. She portrayed this man, clearly destroyed by an almost lethal dose of post-traumatic stress disorder, drugs and liquor. It was a compelling performance.

Anderson, still speaking as this sad middle-aged man, described a particularly difficult episode in his life. 'I had been drinking. I was very loud and belligerent that night and my sister, who is probably the closest person to me, walked off and said that she was never going anywhere with me again. Afterward, I went out to the Dumpster and I threw the bottle in that Dumpster and I said that I was never going to drink anymore. I tried to stop on my own, but the next morning when the liquor store was open I was right there buying another bottle. A lot of times, people – they want off but they have no control. That is what the bondages of Satan do, using alcohol and drugs.'

When she finished this man's monologue, Anderson sang a reprise of the sad song that she'd started with. As she sang, a slide appeared on a screen behind her. Anderson seamlessly switched into a professional voice, with crisp diction and shorn of any accent as she read a re-creation of what a hospital admission note from any of his many hospital admissions might have read. 'Chief complaint – a thirty-four-year African American male brought in by police; a question of a drug overdose.

'The history of the presenting illness: The patient was found unresponsive and brought to ER. He was intubated in the field to protect his airway since he was actively seizing, which caused respiratory depression when he was found. In the ER, the patient was minimally responsive to pain. Per police, he had 3 grams of cocaine in pocket. He has been identified by his driver's licence as Mr R. Johnson whose prior medical records indicate multiple past admissions for drug overdose.'

The students sat in rapt silence throughout the hour-long performance. The contrast between the rich, detailed life portrayed by the young doctor-reporter and the spare, cold language with which it was portrayed in the imagined, but realistic, admission note could not have been stronger. Afterward the students sat in small groups discussing the morning's event. They were moved by the patient's story and horrified by its translation into the coolly impersonal language of medicine.

Angoff sees this as an opportunity to demonstrate what patients see all the time: the cold and depersonalising language and process of medicine. 'I want to remind our students that there's a real person here.' Medical students fall in love with what the doctor's story can do, what medicine can do, she tells me. The morning's performance is there to remind them of what a patient's story can do and how the infatuation can look and sound to the patient they are trying to help.

At the end of the morning Angoff said a few words to the students, summarising what she hoped they have learned. 'You're starting out on the journey across this bridge, this education, and right now you are on the same side as your patients. And as you get halfway over the bridge you'll find yourself changing and the language the patient had and you had is being replaced by this other language, the language of medicine. Their personal story is being replaced by the medical story. And then you find yourself on the other side of that bridge – you're part of the medical culture. When you get there, I want you to hold on to every bit of your old self, your now self. I want you to remember these patients.'

# PART TWO

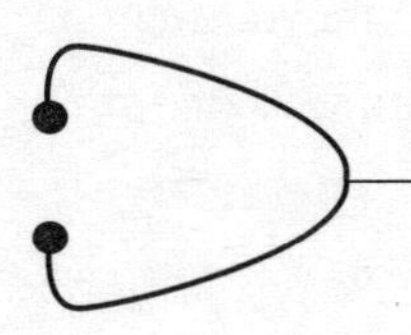

## *High Touch*

## CHAPTER THREE

## *A Vanishing Art*

Here's a story I read not long ago in the *New England Journal of Medicine:*

A man in his fifties comes to an emergency room with excruciating chest pain. A medical student is told to check the blood pressure in both arms. He checks the closer arm and calls out the blood pressure. He moves to the other side of the patient but is unable to find a blood pressure. Worried that this is due to his inexperience rather than a true physical finding, he says nothing. No one notices. Overnight the patient is rushed to the operating room for repair of a tear in the aorta, the vessel that carries blood out of the heart to the rest of the body. He dies on the operating table.

A difference in blood pressure between arms or the loss of blood pressure in one arm is strong evidence of this kind of tear, known as a dissecting aortic aneurysm. The student's failure to speak up about his inability to read the blood pressure on one side of the patient's body prevented the discovery of this evidence.

Here's another story – this one from a colleague of mine:

A middle-aged woman comes to the hospital with a fever and difficulty breathing. She'd been treated for pneumonia a week earlier. In the hospital she's started on powerful intravenous antibiotics. The following day she complains of pain in her back and weakness in her legs. She has a history of chronic back pain and her doctors give her painkillers. They do not examine her. When her fever spikes and her white blood cell count soars, the team gets a CT scan of the chest, looking for something in her lungs that would account for a worsening infection. What they find instead is an abscess on her spinal cord. She is rushed to surgery.

Had the team examined her, they would have found a loss of sensation and reflexes, which would have alerted them to the presence of the spinal cord lesion.

This story was recently presented at Grand Rounds, a high-profile weekly lecture for physicians, at Yale:

A man has a heart attack and is rushed to the hospital, where the blocked coronary artery is reopened. In the ICU, his blood pressure begins to drop; he complains of feeling cold and nauseated. The doctors order intravenous fluids to bring up his dangerously low blood pressure. They do not examine him. When, after several hours, his blood pressure continues to drop, the cardiologist is called and she rushes back. When she examines him she sees that his heart is beating rapidly but is barely audible. The veins in his neck are distended and throbbing. She immediately recognises these as signs that the man has bled into the sac around his heart – a condition known as tamponade. It is a well-known complication of the procedure she'd done just hours before. She rushes him back to the OR and begins draining the blood, which by now completely fills the sac, preventing the heart from beating. Despite her efforts, the man dies on the table. Had the doctors in the ICU examined the patient, rather than paying attention only to the monitors tracking his vital signs, they would have been able to diagnose this potentially reversible complication.

This is another kind of story doctors tell one another in hospital hallways and stairwells – cautionary tales from the pages of our best journals, cases presented at the weekly Grand Rounds or Morbidity and Mortality Conferences, where medical errors are traditionally discussed. These are the tragic stories of patients who worsen and sometimes die because clues that could have and should have been picked up with a simple physical examination were overlooked or ignored. We repeat them to one another as lessons learned – a prayer and talisman. We tell them with sympathy because we fear that any one of us might have been that doctor, that resident, that medical student.

These anecdotes reveal a truth already accepted by most doctors: the physical exam – once our most reliable tool in understanding and diagnosing a sick patient – is dead.

It wasn't a sudden or unanticipated death. The death of the physical exam has been regularly and carefully discussed and documented in hospital hallways and auditoriums and in the pages of medical journals for over twenty years. Editorials and essays have posed once unthinkable questions like: 'Physical diagnosis in the 1990s: Art or artifact?' or 'Has medicine outgrown physical diagnosis?' and 'Must doctors examine patients?' And finally in 2006, the flat announcement of the long-anticipated death was carried in the pages of the *New England Journal of Medicine*. In 'The Demise of the Physical Exam,' Sandeep Jauhar tells the story of that inexperienced medical student – himself – who couldn't find a blood pressure on a man with chest pain and an aortic dissection who dies as a result. It is the tasty opening anecdote in an obituary – not for the patient but for this once valued part of being a doctor.

The physical exam was once the centrepiece of diagnosis. The patient's story and a careful examination would usually suggest a diagnosis, and then tests, when available, could be used to confirm the finding. These days, when confronted with a sick patient, doctors often skip the exam altogether, instead shunting the patient directly to diagnostic imaging or the lab, where doctors can cast a wide net in search of something they might have found more quickly had they but looked. Sometimes a cursory physical examination is attempted but with few expectations as physicians, instead, eagerly await results of a test they hope will tell them the diagnosis.

Many doctors and researchers are troubled by this shift. They complain about the overuse of expensive high-tech tests and decry the decline of the skills needed to conduct an effective physical exam. Yet despite this uneasiness, doctors and even patients increasingly prefer what they perceive to be the certainty of high-technology testing to a low-tech, hands-on examination by a physician.

## *Measuring the Loss of Skills*

In the early 1990s Salvatore Mangione, a physician and researcher at Thomas Jefferson University Medical Center in Philadelphia, began studying how well doctors were able to recognise and interpret common findings on one fundamental component of the physical exam, the examination of the heart. He tested 250 medical students, residents, and postgraduate fellows specialising in cardiology from nine different training programmes. The investigation was straightforward enough: students and doctors were given an hour to listen to twelve important and common heart sounds and answer questions about what they heard.

The results were stunning and controversial. A majority of the medical students could identify only two out of the twelve sounds correctly. The other ten were recognised by only a handful of the students. Surprisingly, the residents did no better. Despite their additional years of experience and training, they were able to correctly identify only the same two examples. Perhaps most disturbing of all, most of the doctors holding a post-residency fellowship in cardiology were unable to identify six out of the twelve sounds.

In a similar test on lung sounds, Mangione again found that students and residents couldn't identify many of the most common and most important sounds of the body. If letter grades were being handed out, all but a handful of these participants would have got a big fat F.

In the years since Mangione first published his studies, editorials and lecturers have bemoaned this loss of skills and warned that if action isn't taken to remedy the problem, we'll end up with teachers who know no more than their students, a case of the blind leading the blind. A recent study suggests that that day has already come. Jasminka Vukanovic-Criley, a physician at Stanford, compared cardiac exam skills of practising physicians to those of medical students and residents. First-year medical students correctly answered just over half of the questions. Graduating medical students were a little better – correctly answering nearly 60 per cent of the questions. But after graduation from medical school, all improvement stopped. Residents, their teacher-physicians on the faculty, and doctors in the surrounding community did no better than graduating medical students.

How did we get here? How can we have generations of doctors who make it through residency and sometimes subspecialty training without improving their skills in the physical exam? Mangione surveyed medical training programmes about their curriculum in these areas and found that only one in four offered structured teaching of basic physical examination skills. Routine observations of trainees performing the physical exam were rarely done. Perhaps, Mangione suggests, doctors don't learn this because programmes don't teach it.

Historically, residency and fellowship programmes rarely taught these skills outright, as a separate course of instruction, because this kind of teaching happened informally, throughout the day, while taking care of patients. At one time, a 'resident' actually lived in the hospital, literally resident, so that he might learn his skills via total immersion, like a Berlitz language class of the body. Part of the total immersion was to pick up the physical examination skills of the older doctors as the resident watched the pros working from room to room.

After every call night, teaching physicians would see each newly admitted patient along with the resident, the interns, and the students. Together they would review the story of the patient's initial presentation and then examine the patient, reviewing significant physical findings noted (or not) by the team. In addition, three times a week, the attending met with residents and medical students for a ninety-minute educational session. During these classes attending physicians were expected to incorporate instruction in the nuances of the physical exam – at the bedside, with the patient.

These types of unstructured, informal teaching sessions, based on the pathology of the patients, were the principal methods of teaching the physical exam along with other aspects of patient care.

Several trends completely unrelated to education have eroded these traditions. First, the rising cost of hospitalisation has focused efforts on shortening patients' time in the hospital. Those with significant heart murmurs, the kind that make good teaching cases, are in and out of the hospital within days. In 1980 the average length of stay in a US hospital was more than a week. In 2004 that had dropped to just over three days. So there is less opportunity to do bedside teaching – a triumph of medical economy that only slowly has been recognised to have come at the expense of education. Patients zip in and out of the hospital too quickly for residents to learn from their exams.

These days, the residents who care for the patients also zip in and out of the hospital. The eighty-hour working week, mandated in 2004 by the Accreditation Council of Graduate Medical Education (ACGME), the organisation that oversees medical education, means that the time that doctors-in-training are allowed to spend in the hospital is limited. Eighty hours may seem like a long working week, but there's plenty to fill it; the amount of work hasn't decreased, only the time available to do it in. What this usually means is that residents spend less time with their patients. In a recent study done at Yale, interns were found to spend less than ten minutes a day with each of their patients.

As an intern, I used to allow two hours to see my patients first thing in the morning, before work rounds when I presented the patient to my resident and the attending. This gave me plenty of time to talk with the patient, examine him, check his labs. With the eighty-hour working week, interns in our programme are not permitted to come into the hospital any earlier than one hour before work rounds. Given the dual demands of patient care and education – which are, after all, the purpose of residency – something had to give. Unfortunately, what's given up is the time doctors spend with the patient.

Our successes in medicine have taken their toll as well. Many diseases are caught early, before the severe consequences are manifested. In the 1990s when I did my training, I was exposed to far fewer types of murmurs and other heart sounds than the generations of doctors who preceded me. Rheumatic heart disease used to be commonplace. In this disease, a strep infection of the throat or skin can cause the immune system to attack the heart, destroying the valves. The unexpected link between this painful but not life-threatening infection and potentially lethal destruction of the valves of the heart was recognised in the early twentieth century. Now physicians routinely check for strep when patients come in with a fever and sore throat. Those with positive tests are treated with antibiotics. The drugs don't do much to shorten the illness or lessen the pain of the infection, but they prevent the development of rheumatic heart disease.

It's been a very successful strategy. Rheumatic fever was one of the most common diseases in America through the 1940s. In 1950 approximately 15,000 people died of rheumatic heart disease; in 2004, that number had dropped to just over 3,200. It's a dramatic decline, but we didn't wipe out the disease, so doctors still have to recognise it when they see it. It's just that now there are a lot fewer patients to learn the symptoms from – an unintended consequence of good medicine.

There are many diseases that are now routinely treated early, often before patients ever need to come into the hospital. It's a success story in medicine and a blessing for patients, but a problem for education based on chance patient encounters in the hospital. The old system of informal teaching, based on learning at the bedside, doesn't work anymore. And medical education has been slow to come up with alternate ways to teach doctors the critical skills necessary for a thorough physical exam.

This loss of skills has resulted in a loss of faith in what the physical exam can do. The official line in medicine is that the physical exam is important. But what you quickly pick up in the 'hidden curriculum' – the values and beliefs of medicine as it's practised – is that the physical exam is mostly a waste of time. On rounds in the hospital, as a student or intern, you might proudly describe a murmur you picked up on exam, but it doesn't take long to realise that it's only the report of the 'echo' (shorthand for echocardiogram

– an ultrasound of the heart) that anyone pays attention to. And because the physical exam is not valued, you soon learn not to pay attention to it and all further learning stops – replaced by the kind of learning you know those who are in charge *will* value. What did the newest high-tech test say? What is the most current research on a particular therapy? These are the questions physicians are now being trained to ask – not the more traditional questions, such as, What did you see when you looked at the patient? What did you feel? What did you hear?

These structural changes in modern medicine – where doctors and their patients zip in and out of the hospital with an ever changing variety of diseases – are expressed at the practical level in this hidden curriculum. But I suspect there is one more reason that the exam has lost its once central position in the evaluation of the patient. In contrast to the cool answers provided by technology, the physical exam feels primitive, intimate – even intrusive. Even when the patient is available and willing, conducting such an exam is psychologically daunting for the physician. It's a truth I learned early in my own medical education.

## *Palpable*

'Do you want to feel my cancer while it's still here?' Joan asked me one wintry afternoon as we sipped coffee in her kitchen. 'You're going to be a doctor. Shouldn't you know what a breast cancer feels like?'

My husband and I were visiting his oldest sister one February weekend in 1993. It was spring break at the Yale Medical School, where I was in the middle of my first year. The week before, Joan had gone for her regular mammogram. As she was getting dressed after the test, the radiologist, an old friend, burst into the room. 'She looked at me and I could tell something was wrong,' Joan told me.

The radiologist arranged for her to see an oncologist, who, in turn, sent her to a surgeon for a biopsy. Our visit caught her before she'd heard the results of the biopsy but well after Joan had accepted the likelihood of the diagnosis.

Joan sighed and tucked a wayward blond curl behind her ear. 'Wouldn't it be helpful to know what to look for? Wouldn't it?' she persisted. After the needle biopsy, she'd located the tiny nodule that was going to change her life and found herself touching it several times a day, the way you sometimes can't stop fingering a painful sore or replaying a difficult conversation in your head – acting on some need to remember where the pain was coming from.

I didn't know what to say. I had no idea what a breast cancer would feel like and she was right – it would be useful for me to know. And I was wildly curious.

But I knew immediately that I couldn't do it. Touching my sister-in-law's breast was inconceivable. Joanie was able to imagine me in the role of a physician – a group given permission to ignore the traditional zones of privacy when necessary. But it wasn't a mantle I was ready to put on. At that point in my training, I had not yet examined anyone. Until that moment I hadn't really envisioned how strange and unnatural it would to be to violate the zone of privacy each of us occupies. I couldn't touch my sister-in-law. In fact, I wasn't sure I could touch anyone.

The act of placing your hand upon another's body is, in many ways, the hallmark of the physician. And yet, though simple, it is an act riddled with complications. Who are the people we touch in our lives? Our lovers, certainly; our children, naturally. And as a sandwich generation, perhaps even our parents, eventually. No one else. I don't count the hug and cheek-peck hello, the hand on the shoulder, the slap on the back. This is touch as a form of communication – it speaks of fellowship and affection, support and concern. This type of physical contact lies well within our expectations of social intercourse. It is by convention brief, by practice unobtrusive. A hug or touch that lasts a little too long or is a little too close sets off alarms because we understand the rules of social conduct.

In medicine, at the bedside, on the examination table, we touch those we care for – but it's a different form of touch, and a different kind of care. Medicine requires intimacy but one characterised by an intellectual and emotional distance. You don't expect your friends and loved ones to assess you with a knowing and impartial eye. We allow them to occupy an intimate space physically and emotionally because we know they see us through a filter of love.

The intimacy of the physical exam is far removed from that between friends and family. In the physical exam, that filter is gone. Doctor and patient are often strangers to each other. It can be uncomfortable – for the patient, and often for the doctor as well. And there is, at the heart of this sometimes awkward intimacy, a fiduciary relationship, an implicit bargain: the patient will let the doctor see him and touch him and in return the doctor will share her knowledge for the benefit of the patient. When Joan had her cancer, I knew I wasn't ready to live up to my end of the deal. I had nothing to offer: I knew a lot of anatomy, some cell biology, a good deal of genetics, but I didn't know anything about medicine. Not then.

Moreover, I didn't know how to do it. Literally. I hadn't been taught. That was something I would learn in my second year. Perhaps even more important, I hadn't yet learned how to occupy that permitted space between physical intimacy and intellectual distance that is fundamental to touching as a doctor. That part isn't on the written curriculum; there weren't any lectures on it (or at least not in my medical school), and yet you can't be a doctor if you don't learn how to negotiate this deeply personal territory. Medicine – to the extent that it can be called a science – is a sensual science, one in which we collect data about a patient through touch and the other senses according to a systematic method in order to make a diagnosis. Most patients are willing to be touched by their doctor. They expect it. I certainly expected to touch patients. But, as I realised that afternoon in my sister-in-law's kitchen, first you have to learn how.

In medical school, starting with anatomy class, doctors are taught to understand the body by taking it apart, one piece at a time. What you walk away with, at the minimum, is an uncanny ability to objectify the hell out of even the most intimate body parts. For anyone else, this might be considered disrespectful, but for doctors, a clinical and objective view of, say, a female breast offers us the chance to see it isolated from its other, often sexual, contexts. We are taught to handle a breast as a separate object.

And so when you examine a breast, you notice that the smooth skin and soft layer of fat beneath give way under your fingers to reveal highly organised, dense layers of glandular tissue below. Beneath the skin, trapped by the investigating hands, the breast is so much more orderly than the wildly mobile appendage it appears. I learned how to examine the breast in the middle of my second year. A patient-instructor – a layperson trained in the techniques of this exam – walked me and the three other students who made up my physical exam group through the methodical examination using her own breasts as the models on which we learned.

As the class began, I felt again that same discomfort I'd felt in Joanie's kitchen. We were four medical students, dressed in our still creased short white coats, our shiny name tags pinned to our lapels, and brand-new stethoscopes folded into our pockets, trying hard to appear relaxed as we sat in a semicircle around a half-naked middle-aged woman. The teacher sat comfortably on the exam table. The robe she had worn when we entered the room was pushed down around her waist to reveal the subject of this class, her breasts. I tried to relax my face – to at least appear at ease. I wasn't quite sure where to look. I pulled out a notebook to take notes as she talked about the exam. I could feel the tension emanating from the students on either side of me. None of us said anything at all. Greg, the earnest, well-meaning liberal from New York's Upper West Side, appeared to be studying his shoes. Lillian, the exuberant effusive life force in our class, fidgeted quietly with her hair. No one made eye contact with the teacher or one another. The four of us silently struggled to figure out a way to manage our discomfort. I knew then that those skills were part of what the class was supposed to teach us.

'First you inspect,' she told us. 'The breasts should be symmetric.' She raised her hands above her head, and her breast stretched upward as well. Then she placed her hands down on her hips, spreading her elbows wide to stretch the chest muscles and the breasts. 'This gives you a chance to see any abnormalities that affect the shape of the breast or the texture of the skin.'

As she ran through her orderly presentation the tension in the room began to fade. Her matter-of-fact demonstration communicated an ease with her role as a patient and her expectations of us as doctors. I realised later that she was not only teaching the basics of the breast exam, she was demonstrating a technique we could use in our own encounters to bring a naturalness to the awkward physical intimacy between a doctor and a patient.

'I'm going to demonstrate how to examine my breasts and then each of you will do it.'

We crowded around the table as she lay with one hand over her head and the other performing the examination.

'Start at the midline. Use the pads of your fingers. Press them lightly on the breast and make a circle. I like to anchor my fingers to each spot and move the skin along with my fingers so that I know that I'm examining the same spot. You do that three times at each spot, each time applying just a little more pressure so that you can feel all the different structures under the skin.'

We watched with interest as she repeated this motion up and down her chest in overlapping stripes from her sternum all the way over to her side and up into the dome of her underarm.

'Now you try it.'

As I stepped up to the exam table I could feel my discomfort reemerging. I rubbed my suddenly cold hands together, trying to coax blood into icy fingers, then placed them on her chest. Her skin was warm and I could smell the fragrance from the soap or deodorant she'd used that morning. Her professional but casual tone kept me focused on the medicine at play here and not the intimate zone that she and I had suddenly entered.

'Follow the clavicle over to where it meets the sternum,' she instructed me. Her voice was patient, comfortable, completely relaxed. I moved my fingers self-consciously, in an awkward circle over the skin, bone, and cartilage. Next to the sternum a thin film of fat overlies smooth, tough muscle and ribs. Further down the chest the layer of fat gets thicker at the outer region of the prominence we recognise as breast. It wasn't until I started the second stripe that I encountered the fingerlike irregular densities of the glands themselves, pointing inward to the nipple like the spokes of a wheel. As I made my way across the breast, images from my anatomy book depicting these structures I could only feel flooded my mind, like aerial photographs providing landmarks and explanations of the terrain beneath my fingers. The area directly under the nipple dips like a soft well into this dense tissue; I could picture the ducts, too fine to be felt. Below that, I felt a saucer, a hockey puck of thick consolidated gland too closely packed to be individually distinguished.

As I worked my way across her breasts, she offered advice and encouragement.

'You can use more pressure than that if you need to. It doesn't hurt me. Use your other hand to stabilise the breast.'

I covered her chest with the lines of circles, working to make sure I could feel every structure below the skin from as many angles as possible. I thanked the instructor and gratefully stepped back from the exam table as the next student stepped forward. I watched as she coaxed and encouraged my three classmates through the exam and reviewed the process in my head from the safety of my chair.

A couple of years ago I moved my practice from one office to another. As I reviewed the charts of my relocated patients, transferring data from old to new, I noticed that although I had done a pretty good job in making sure that my patients got their recommended screening tests, I hadn't done nearly as well on the hands-on component. Women should have a breast and pelvic exam performed annually, I was taught. Men over fifty should have a yearly rectal exam to look for prostate cancer. I saw that my adherence to those guidelines was pretty spotty. I was surprised by this oversight, but the trend was too strong to deny.

I puzzled over this. How could this happen? Some of it was a systems problem. In my old office there was no simple way to keep track of routine exams. To find the last exam I'd have to page through the last year's worth of visits to see where I'd documented the results. And yet regular cholesterol tests were there. My patients over fifty had colonoscopies ordered or at least discussed. No, it was the breast exams, pelvic exams, and prostate exams that were missing. And I realised that despite the years of practice and the mastery of technique, I still found these exams uncomfortable to perform.

On some level, I was still that medical student, reluctant to touch another person's private places.

I'm not alone in this. There's not a lot of data on this issue, but what's there suggests that more of us are sending our patients for the screening test and dispensing with the hands-on component. In a study published in 2002, of the 1,100 women who went for annual mammograms in one facility over the course of a year, only half reported having had a breast exam done by their physician – ever. And while rates of mammography have increased over the past twenty years, rates of physician breast exam have declined.

Is that all due to the awkward intimacy of the exam? Probably not, though research has shown that it plays a role. Instead, the development of newer and better technologies – the mammogram, ultrasound, most recently the MRI – has caused doctors to doubt the value of what their hands can tell them. Why deal with your own embarrassment, the possibility of patient embarrassment, and the difficulty of interpreting the fuzzy pictures generated by touch when a study can show you the inner structures of the body with more precision and accuracy?

Why indeed? I'll explore some of the answers to this increasingly urgent question in the next chapter.

# CHAPTER FOUR

## *What Only the Exam Can Show*

As the skills required for an expert physical exam have become more and more rare, both among medical students and among practising physicians, what has been lost? Among doctors, this is a topic of passionate debate.

On one side are those who argue that the demise of the physical exam is a natural consequence of progress. They say that the exam is just a charming remnant of a bygone era – like cupping (attaching warmed cups to the skin until blisters are formed) or bleeding or mustard plasters for colds – now replaced by an ever enlarging menu of technologies that provide better information with greater efficiency and accuracy. Affection for this discredited practice is characterised as pointless and sentimental.

On the other side are the romantics: doctors who see the physical exam as part of the long tradition of caring in medicine and cherish the profound connection between doctor and patient when linked by a well-placed hand and a warm heart. They see those who think otherwise as soulless technicians.

In the middle are the rest of us who simply want to understand what's been lost. How large a role did the physical exam once play in making a diagnosis? What are we missing in the modern version of medicine that somehow seeks to manage without it?

---

Steven McGee, a mild-mannered man with a serious face, an FM radio voice, and a scholarly passion for the physical exam, has blazed a rational trail deep into that middle ground. As an internist and a professor of medicine at the University of Washington, he embraces technology but also believes that the physical exam has uses that machines cannot replicate. McGee's research is an outgrowth of his own experiences in medicine, and his book, *Evidence Based Physical Diagnosis,* outlines the evidence for the utility of the physical exam in the age of high technology.

When I spoke with McGee about his work, he was eager to tell me about examples from his own experience of medicine that proved to him the fundamental importance of examining the patient. He recalled a particularly dramatic case that had occurred just a few weeks before we spoke.

McGee and his team of residents and medical students were called to see a patient on a surgical floor. The patient had come to the hospital for the excision of a skin cancer on his ear. That morning he'd developed severe abdominal pain, and the plastic surgeons had asked them to help figure out what was going on.

Michael Killian, a thin elderly man, lay on the bed with his eyes wide open, moving restlessly as if he couldn't find a comfortable position. He muttered incoherently as he shifted awkwardly across the bed.

The resident introduced himself to the distraught patient and immediately began asking questions. 'I don't know. I don't know. I don't know,' was his only answer. It quickly became clear that the elderly man was too confused to provide any details about his pain. He could tell them his name. But he didn't seem to know that he was in the hospital or why. All he could say was that he hurt. When the resident asked if he had pain in his belly, he started his litany once more: I don't know, I don't know.

His skin was pale and littered with scaly patches of red, evidence that he'd spent too many hours in the sun. The ear that had brought him to the hospital in the first place was enlarged and distorted by a raised red and scaly lesion at the tip. His unshaved cheeks were gaunt, his cheekbones sharply defined, his eyes seemed focused on something in the room no one else could see. A fringe of white hair was well cut but uncombed. His skin was cool and damp with sweat. It was difficult to examine him because of the constant restless movement. His heart was fast but regular. So was his breathing. When the resident moved to examine the patient's abdomen, he jerked away. 'No. No. No. Don't touch me.' The distant eyes were now back in the room, glaring at the young doctor. The patient waved his arms in a way that suggested that no means no. The doctor quickly pulled back.

'No. No. No.'

The resident leaned down and began to speak in a quiet voice to the distressed man. 'I know you are in pain and I want to help you. But in order to help you I need to touch your stomach. I won't hurt you.' The soothing tone eventually quieted the suffering man, though he continued to shift his position on the bed, as if the soft mattress had been replaced by a bed of nails.

As the resident reassured the confused and frightened man, McGee gently placed his hand on the upper left side of the man's abdomen. He felt an unexpected resistance in the normally soft region of the belly and quiet steady pulsations. He placed his other hand over the man's navel. A soft mass throbbed beneath his fingers, pushing his fingers away to the right. And that told him everything he needed to know.

'Call the surgeons,' McGee told the resident. 'This man needs to go to the OR. He's got a rupturing aortic aneurysm.'

The aorta is the vessel that carries blood from the heart to the rest of the body. Patients with hardening of the arteries and high blood pressure – like this man – can develop areas of weakness in the normally thick muscular tubing, and the stress of this high-pressure system can cause these weak spots to balloon outward, forming a pulsating bulge in the abdomen. When the balloon gets large enough, the muscle wall becomes dangerously thin and it's at risk for bursting. The excruciating pain and the restless movement were classic for a tear in the now delicate muscle wall, and the huge pulsating mass clinched the diagnosis. Three quarters of all patients who suffer this dire event die either on the operating table or on their way there.

The vascular surgeons were paged and the patient was taken to the OR, stopping only briefly at the CT scanner to verify the diagnosis. Defying the odds, Mr Killian survived the surgery, his life saved by a simple touch.

As compelling as any individual case may be, in medicine, if you want proof you need studies. And McGee has spent his career investigating and tabulating the accuracy of individual components of that endangered art, the physical exam. His results have managed to anger folks on both sides of the debate. Some well-known, frequently taught parts of the physical exam have turned out to be virtually worthless – listening to the lungs will rarely help a physician decide if a patient has pneumonia. Others, when done well, have shown themselves to be as solid and reliable as the tests we use to confirm our diagnoses. In the hands of experts, a cardiac exam can identify problems in the valves of the heart almost as well as the echocardiogram. It's essential to know how well each of these individual tests performs.

But this research still leaves the big question unanswered: is there any evidence that this old-fashioned practice really makes a *difference* in how patients do? There is surprisingly little research on this. Several now classic studies done in the 1960s and 1970s tried to assess which tools are most useful in helping doctors make a diagnosis. In these studies the most important tool was the simplest – doctors were able to correctly diagnose patients' illnesses in most cases just by talking. The patient's story contained the diagnostic tip-off up to 70 per cent of the time. Doctors are told repeatedly in medical school to listen to patients and they will tell you what they have. These studies prove the wisdom of this advice.

But what about the physical exam? In these same studies, when you looked at just those patients whose story didn't provide the answer, the physical exam led to the right diagnosis about half the time. High-tech testing showed the way in the remaining cases.

Of course, testing has changed a lot since those studies were done. A more recent study, done by Brendan Reilly, a head of clinical medicine at Weill Cornell Medical Center, looked at this question in a different way. Reilly was asked by one of the residents he teaches how important the physical exam was in making a diagnosis. Reilly searched the medical literature for an answer. When he couldn't find a good answer, he designed his own study.

In a teaching service like his, patients are seen first by the internal medicine residents and then are examined and evaluated separately by the attending physician. The residents and the attending swap the information they collected independently to figure out a diagnosis and care plan. Reilly reviewed the charts of all the patients he had admitted to the hospital with his team over the previous six weeks, looking for any case where something he found on the physical exam had changed the diagnosis and the treatment of patients under his team's care.

The findings were pretty impressive. A careful physical exam changed the patient's diagnosis and treatment in twenty-six out of one hundred cases – one in four patients. And in almost half of these cases, had Reilly not discovered the correct diagnosis on exam, it would not have been found by 'reasonable testing' – that is, testing that would have been ordered if these physical findings had not been discovered. In those cases, the correct diagnosis would have only become apparent when the disease progressed and the patient worsened.

These were important discoveries. In one striking case, a patient who was admitted to the hospital for difficulty breathing was thought to have a tumour in his chest, picked up on his admission X-ray. He had been scheduled for a biopsy of the mass. When Reilly examined the patient, he found a loud heart murmur. Based on the location and timing of the abnormal sound, he realised the noise was caused by an obstruction in one of the valves of the heart. The blockage was causing the vessels leading up to the valve to enlarge with the excess blood – the way traffic backs up when construction or an accident narrows a busy highway. The 'mass' seen in the chest X-ray was actually the blood-engorged vessels. The biopsy was cancelled and the patient was referred for the surgical repair of his valve.

Another patient had a fever, but no source of infection had been found. He was being treated with intravenous antibiotics. Reilly noticed that one of the patient's toes was discoloured in a way that suggested the toe had been cut off from the body's blood supply and had become infected. Surgery was consulted and the toe was amputated. The fever disappeared along with the toe.

This handful of studies suggests that a thorough physical examination can play a critical role in making a timely diagnosis – a role that cannot be duplicated by even the sophisticated tests we now have available.

One of the ironies of our technology-laden age is that many of the time-

and labour-saving devices that have crept into our daily lives often save neither. Most computer desktops include a virtual notepad. Is it any better than the actual notepad kept in your pocket? A calculator can be essential for performing complex functions, but does it save time when all you really need to do is add, subtract, or multiply a few numbers?

In the same way, medical testing is one way to come up with a diagnosis, but sometimes – and if Brendan Reilly is right, up to 25 per cent of the time – you can get the right answer by simply examining the patient.

This is not to say that a physical exam can substitute for testing. With the tests we now have at our disposal, we can diagnose diseases that in another era, not so long ago, could be identified only at autopsy. But the physical examination can direct the doctor's thinking and narrow the choice of tests to those most likely to provide useful answers – saving time, saving money, and sometimes even saving lives.

## *The Language the Body Speaks*

The experience of being ill can be like waking up in a foreign country. Life, as you formerly knew it, is put on hold while you travel through this other world as unknown as it is unexpected. When I see patients in the hospital or in my office who are suddenly, surprisingly ill, what they really want to know is 'What is wrong with me?' They want a road map that will help them manage their new surroundings. The ability to give this unnerving and unfamiliar place a name, to know it – on some level – restores a measure of control, independent of whether that diagnosis comes attached to a cure. Because, even today, a diagnosis is frequently all a good doctor has to offer.

That was certainly the case with Gayle Delacroix, a fifty-eight-year-old retired soccer coach and gym teacher who came to the small community hospital in Connecticut I work in with a puzzling illness.

It was in the late summer of 2003 and Gayle and her longtime partner, Kathy James, were on their way home from a two-month camping trek across the US – driving, biking, and hiking from northern Connecticut as far west as the mountains of Colorado. They'd planned to end up in their own beds by the weekend. It had been a great summer, until one night, when Gayle was awakened by an excruciating pain across her lower back. The pain was sharp. Stabbing. Unbearable.

Gayle woke her partner: 'Something's wrong with me,' she told her. In the flickering glare of the flashlight Kathy saw that Gayle's face was slick with sweat, tense with pain. Though the summer night was cool in the mountains, her skin was hot and Kathy didn't need a thermometer to know that her partner had a fever.

Her head hurt, Gayle told her. And she felt hot and cold at the same time. But worst of all, she had this intense pain across the lowest part of her back. It had that precise yet elusive quality of an ice cream headache. Sharp needles of electricity flashed down the back of her legs every now and then, but the back pain was persistent, gnawing. Her teeth chattered as she spoke. Her body shook with wracking chills.

Kathy realised that Gayle needed a doctor. She dressed and quickly stuffed her sleeping bag into a sack. Helping Gayle out of the tent and onto the stump they'd used that evening for a table, she packed up their gear and hurried down the trail to the car. Then she returned to help her partner down the rough track.

They drove an hour through the back roads of West Virginia to Maryland. Another hour to an exit marked with the white H promising a hospital ahead. The ER doctor was practically a kid. Tall, wiry, with stylish glasses and a rumpled scrub shirt over blue jeans, he looked like he'd just crawled out of bed. He helped Gayle sit up and quickly examined her back.

He offered a diagnosis and some reassurance.

'I don't think the fever and the back pain are related,' he told them. 'I think the back and leg pain is sciatica. And the fever – who knows? Some virus, probably.' He gave Gayle some ibuprofen and a muscle relaxer for her back. When Kathy – angered at the breezy exam and unconvinced by his diagnosis – brought up the possibility of Lyme disease ('We've been camping, for God's sake'), he dutifully wrote out a prescription for doxycycline, the antibiotic of choice for this disease.

Kathy was worried – she was a physiotherapist. She had seen lots of sciatica but none this bad. And this fever? Hard to believe they weren't related. Gayle, on the other hand, was relieved by the reassuring diagnoses. She had never been sick and wasn't ready to start now. After leaving the hospital they drove until dawn, then checked into a roadside motel and caught up on the sleep they'd missed. They slept soundly – Gayle with the help of the ibuprofen, the muscle relaxer, and, at Kathy's insistence, the doxycycline. When they awoke it was late afternoon.

Gayle sat up. She felt a little better, though her legs were strangely heavy as she swung them to the floor. When she tried to stand, they buckled beneath her and she fell back, helpless, onto the bed.

'My legs aren't working, Kathy. I can't walk.' Gayle's voice was high-pitched and terrified. 'I can't walk,' she repeated.

Kathy's heart began to race. She knew it. There really was something wrong. They weren't far from Baltimore – maybe there? No, Gayle insisted. She wanted to go home.

They were at least five or six hours from the small Connecticut city they lived in. Kathy drove as fast as she could directly to their local hospital. 'It was the longest five hours of my life,' she told me later.

'Stay here,' she instructed her partner and disappeared into the emergency room. She returned a few minutes later with a couple of EMTs – emergency medical technicians – and a wheelchair. The three of them helped the now crippled woman out of the car and hurried her into the ER.

Dr Parvin Zawahir, a first-year resident, was the doctor on call that night. She quickly reviewed the thin chart that documented the patient's time in the ER. A fever of 101. Weakness. The blood work already done didn't show much – the white blood cell count wasn't elevated. Chemistry was normal. Liver – normal.

She found the patient's curtained-off cubicle, introduced herself, and began the familiar process of taking a history. It had started five days ago, Gayle told her. She had a stomach ache and some diarrhoea. She figured it was a touch of food poisoning and didn't think much of it. Two days later she'd developed a rash on her neck. It didn't itch or hurt and she hadn't even noticed it until Kathy pointed it out. She thought at first it might have been a spot rubbed raw by the strap of her bicycling helmet, but the next day it had spread to her legs and stomach. Then yesterday, she'd felt tired after shooting a few baskets – not her normal stamina. But she hadn't actually felt sick until that pain woke her up almost twenty-four hours ago.

Any bites? Zawahir asked. Gayle nodded. Lots. She'd had plenty of mosquito bites. Didn't recall any tick bites. She hadn't been around anyone who was sick. No pets. She didn't smoke – never had. She didn't drink or use drugs.

The young doctor looked closely at the rash. It was faint but covered much of her body. It was made up of dozens of small, slightly raised, slightly red bumps.

Her back looked normal enough and had no tenderness. The rest of the exam was unremarkable until she got to the patient's legs. Gayle was able to wiggle her toes and move her feet forward and backward. But she couldn't lift her legs – at least not the left one. Zawahir sat down at the desk and started on her admission note. How was she to put all this together? Was this a problem of the muscles? That was the only part of the exam that was abnormal. Or was it the nerves that empowered the muscles? The kind of pain the patient described – with the electric charges down her leg – certainly sounded a lot like the sciatica the Maryland ER doctor had thought it was. But Zawahir couldn't believe that the fever and pain were separate problems. That didn't make sense. They started at the same time. No, they had to be linked.

Infection seemed most likely. Being outdoors for all that time, she was a perfect candidate for Lyme disease. On the other hand, the patient had been in Colorado and West Virginia and a dozen points between – was there Lyme disease in these places? What about Rocky Mountain spotted fever? That was also carried by ticks and characterised by a fever and a rash. And it could be deadly.

Could it be a mosquito-borne illness? In Connecticut, every summer there was a big scare for Eastern equine encephalitis. Though she didn't know how many cases of this disease there were in a year, she'd read that it was frequently fatal. What other viruses could do this? Could this be West Nile virus? Herpes encephalitis? She wasn't sure. She'd never seen any of these illnesses.

She would need to do a spinal tap to see if the lab could find any bacteria or evidence of infection in the fluid. And she would send off for more blood tests as well. An MRI would show if there was an infection in or near the spinal cord. She would start her on high-dose antibiotics – one that would cover both Lyme and Rocky Mountain spotted fever. And she'd like to get an infectious disease consult. Maybe a specialist could help her figure this case out.

Although she'd taken care of sicker patients, the intern was worried about the near paralysis of the patient's legs. If you catch a neurological injury early enough you can sometimes reverse the damage. If not, this youthful, active woman could be crippled for life.

After rounds the next morning, Zawahir sought out Dr Majid Sadigh, an infectious disease expert in the hospital and one of the smartest doctors she knew. Every doctor knows someone like this – the guy you go to when you're stumped. Or worried. Or scared. In every hospital or community of physicians, there is always that one doctor whose clinical acumen and breadth of knowledge seem far greater than anyone else's. There is no list of such names or awards given for this honour. It's simply word of mouth among physicians. In central Connecticut, Sadigh was one of those doctors.

Majid Sadigh had trained in infectious disease in his homeland of Iran. In 1979, not long after Sadigh had completed his training, Mohammed Reza Pahlavi, the US-supported monarch (known here as the Shah of Iran), was overthrown in a religious revolution and Sadigh and his family were forced to flee. He ended up in Waterbury, Connecticut. In order to practise medicine in the US, all foreign-trained physicians have to complete a residency here, regardless of their previous experience. The programme Sadigh was accepted into was small but widely respected for the high quality of its teaching. Sadigh's skills were so impressive that by the end of the first year of what is normally a three-year programme, he was made chief resident. The following year, he joined the faculty at Yale Medical School and has been there ever since.

From the first days of his residency, Sadigh realised that he had a skill almost unknown in the US: he understood the techniques and the value of the physical examination. In Iran even simple tests are often unavailable. In this setting a physician must rely on the patient's story and physical exam to make a diagnosis. 'The body is there, filled with so much, so much to tell you. But if you do not speak the language, you will be deaf to its secrets. My job,' he told me, 'is to teach our residents this important language.'

Zawahir briefly laid out the case for Sadigh, then took him to the patient. The young doctor watched with interest as Sadigh spoke to Gayle and Kathy. He sat down next to the bed and began to question the two women about what had happened. Then he carefully examined Gayle, paying special attention to the affected left leg. He elevated both heels, cupping them in his palms a couple of inches above the sheets.

'Lift your right leg,' he instructed. As she struggled to raise the weakened right leg, the paralysed left leg sank a bit, but not low enough to touch the sheets.

'Now lift the left.' Gayle bit her lip as she strained to elevate the partially paralysed leg. As she worked, the right heel sank down to the bed as she recruited the strength in her hips to raise the leg. The left leg never budged. Replacing her legs on the bed, he tested the strength in her lower legs.

'Push against my hand with your feet like you are stepping on the gas.' The right foot flexed forward; the left barely moved. He touched her gently on both legs.

'Can you feel this?' She nodded. 'Is it the same on both legs?' Again she nodded. He worked his way up her legs. Sensation was normal. He lifted her left knee with one hand and struck it with a rubber arrowhead hammer. Nothing. He repeated the move on the right. The leg jerked and swung upward. He tried again on the left and again there was no response at all.

He stared at the left leg, then called Zawahir over. 'Look at this,' he said, pointing to the patient's leg. Tiny patches of skin on Gayle's leg appeared to be moving, jerking, twisting. There was no movement of the leg itself – just the skin and the muscles of the thigh. Small groups of muscles were contracting spontaneously, independently. It looked as if there were little worms inching along under the skin.

'Fasciculations,' said Sadigh in his soft accented voice; little uncoordinated bursts of activity from a group of muscle fibres powered by a single nerve fibre. He knew he had found an important clue.

Outside the room, Sadigh reviewed what he thought were the important characteristics of the patient and her illness: First, she had been very healthy until now and had spent a lot of time outdoors. She had a profound weakness that affected both legs, but one much more than the other. It was only the thigh and hip muscles that were involved – the muscles of the lower leg and upper body were spared. Only the nerves that power the muscles were affected. Sensation, which is carried on different nerve fibres and connects to a different part of the spinal cord, was normal. And she had fasciculations. Those little muscle jerks were the clincher. The fasciculations and the sparing of sensation suggested that a single type of cell in the spinal cord was affected: the cells that control the muscles of the body, known as the anterior horn cells – a description based on where they are located in the spinal cord.

'I've seen this before – but not so much in this country. This is what polio looks like,' he said – then added: 'But I do not think this is polio.' There is another disease, he explained, a disease new to the US. A disease that can look just like polio. A disease that can cause the same devastating paralysis. He paused. 'I think she has the West Nile virus.'

West Nile had burst into the news four years earlier in the summer of 1999, when it ravaged a small community in Queens, New York. It was a disease well known in Africa, where it originated, and localised epidemics had been reported throughout Europe and parts of Russia, but until that summer, it had never been seen in the United States. The distinctive presentation of the disease – with its polio-like paralysis and its preference for those over fifty – had helped the Health Department doctors in New York recognise it as a new entity and move rapidly and aggressively to contain the epidemic. Nevertheless, sixty-two people were hospitalised with the virus that summer; seven of them – all over fifty – had died. Despite aggressive measures to wipe out the mosquitoes that spread the disease, by 2003 cases had been reported in every state in the continental United States.

Sadigh remembered the events of the summer of 1999 clearly. The polio-like quality of the disease had been much discussed at the time. Seeing Delacroix, Sadigh was certain this is what she had. A sample of Gayle Delacroix's spinal fluid had to be sent to the state lab in Hartford to confirm the diagnosis. It would be days – maybe weeks – before the results would be available. In the meantime they would make sure that it wasn't some other entity that they would need to treat.

After discussing the likelihood of West Nile virus with Dr Sadigh, Zawahir returned to the patient's bedside to tell her the news. Gayle and Kathy had heard about West Nile virus. Who in Connecticut had not? But they didn't know much about it. Zawahir made the parallel to polio that Sadigh had made. When she heard that, the patient's eyes filled with tears. The very word brought up images of children in iron lungs or walking with metal braces and crutches. Was that her future? Zawahir tried to reassure her but she didn't know. This was one of the first cases seen in the state. They'd simply have to wait and see what happened.

'The hardest part was not knowing what was going on or where this would take me,' Gayle told me. The diagnosis of West Nile virus wasn't reassuring, but for someone relatively young and exceptionally healthy it was survivable. She and her partner found themselves in a whole new world. It wasn't where they wanted to be, but it was where they were, and so they threw themselves into the work of learning a new language, mastering a new landscape.

Kathy read up on West Nile virus and polio, hungry for strategies to help her partner fight back. By her third day in the hospital, though still febrile and weak, Gayle insisted on trying to get out of bed and stand. She did it, though she needed help. By the end of the week she had taken a few unsteady steps braced with a walker and monitored by the physiotherapist. Meanwhile, the test results slowly trickled in. It wasn't Lyme; it wasn't Rocky Mountain spotted fever. It wasn't tuberculosis, sarcoidosis, syphilis, or HIV.

Antibiotics given in the hope of a treatable infection were stopped. Finally they received the confirmation of what they already knew. She had been infected with the West Nile virus.

'We hoped against hope that it wasn't West Nile, but the doctors seemed pretty sure right from the start,' Gayle told me. Just knowing what she was up against – as scary as it was – was unexpectedly comforting and gave her a direction to focus her considerable energy to get well.

## *No Time for a Physical*

In the case of Gayle Delacroix and the West Nile virus, the physical exam led directly to an extraordinary diagnosis. More commonly, the physical exam can provide not a diagnosis but an essential clue to direct further testing – a shortcut to the right answer. Ordering a slew of studies to evaluate a patient might get you the answer eventually, but time is often short in the care of a very sick patient. In many cases a careful exam can focus the search and help the physician find the problem faster. Where such an advantage would be most helpful, naturally, is among those patients who are critically ill. But even here – maybe especially here – the physical exam is becoming as obsolete as the doctor's black bag.

The sicker the patient, the greater the temptation to skip the fundamentals – like the physical examination – and to rely on the available technology to provide us with answers. It's a temptation that can sometimes prove fatal – as Charlie Jackson almost discovered.

For most of his adult life Charlie Jackson didn't go to doctors. That changed when he had a massive stroke at age sixty-two. The stroke rendered his right leg and arm nearly motionless, his face crooked, and his speech slurred. Still, his beautiful cockeyed smile and gallant manner – he frequently showed up for his appointments toting a basket of peaches or a bag of pecans from back home in his native Carolina – made him a favourite at our office. He had been doing well, so I was shocked when I got a call from the staff saying that Charlie was dying.

He'd come to the office for a regular follow-up appointment with Sue, our nurse-practitioner. As soon as she saw him that morning, she knew that there was something very wrong. His walk, always a little ungainly after his stroke, was barely a shuffle. His slender frame was bent over his walker as if he couldn't hold himself up.

'What's the matter, Charlie?' she asked as she hurried to his side. 'I . . . can't . . . walk.' He choked out the words. His voice was strange in a new way, too – as if he were speaking in slow motion. She reached down and felt his pulse. It was slow – very slow. Too slow to keep even this slender reed of a man alive. She didn't do any more of an examination. She knew he needed to be in a hospital.

The EMT team burst through the emergency room doors, pushing Charlie into the throng of the crowded room. The triage nurse directed them straight into an empty cubicle as they barked out what they knew. 'Sixty-four-year-old man . . . history of a stroke . . . complaints of weakness and belly pain.' His heart was slow, they reported; his blood pressure too low to be measured. The monitor showed a heart rate in the twenties – normal is over sixty. Dr Ralph Warner strode in and quickly assessed the situation. 'Get me an amp of atropine,' he snapped, calling for the medicine used to speed up the heart.

After injecting the medicine, he watched as the monitor continued its flat yellow line, broken far too rarely by the spike indicating another heartbeat. But slowly the patient's heart rate and blood pressure began to rise.

With the usual chaos of the emergency room boiling around them, Warner forced himself to sit and focus as Charlie described his symptoms. It had started the night before, he told the doctor in his new, strange slur. He felt weak, could barely move. That morning his stomach began to ache. Any chest pain? Warner broke in. Shortness of breath? Fever or chills? Vomiting? The patient shook his head no. He was taking medications to lower his blood pressure and cholesterol. He had not smoked or drunk alcohol since his stroke. A brief exam showed Warner the results of the stroke but he saw nothing else.

Why was his heart beating so slowly? the doctor wondered. Had he taken too much of one of his medications? Had he suffered a heart attack that affected the natural pacemaker in his heart? The EKG, although abnormal, didn't suggest a heart attack. Warner called the cardiologist, who rushed in to place a temporary pacemaker. Charlie was being prepped for this potentially life-saving treatment when the lab called with part of the answer.

Blood work done in the emergency room showed that the patient's kidneys weren't working. And his potassium – an essential element in body chemistry, regulated by the kidneys – was dangerously high. Potassium controls how easily a cell responds to the body's commands. Too little potassium, and the cells overreact to any stimulation; too much, and the body slows down. If the elevated potassium was slowing his heart, then getting rid of the mineral would allow his heart to pump at a normal rate. The patient was given a medicine to get the potassium out of his system and then transferred to the ICU for monitoring.

If the potassium was high because of his kidney failure, what had caused his kidneys to fail? Dr Peter Sands, the intern on call in the ICU, gnawed at this question as he reviewed the chart and results of all the tests that had been done. It wasn't a drug error. The patient's medication box showed the correct number of pills. And it hadn't been a heart attack; a blood test proved that. Sands looked for the results of the urinalysis to see if there was any clue there but he couldn't find it. Somehow no one had sent any urine to the lab. Were his kidneys too damaged to produce urine? That would be critical to know.

Sands asked the nurse to get some urine from the patient. She returned empty-handed. The patient couldn't urinate; he told her he hadn't been able to since the night before. The nurse hadn't been able to insert a Foley catheter, a rubber tube that is passed through the urethra into the bladder to collect urine. Was something blocking the urethra? A urology resident finally managed to get a catheter into the bladder and immediately urine gushed out of the tube – nearly half a gallon of it. A full bladder comfortably holds a little over a cup of urine. Charlie's bladder had held just under eight. The urology resident looked at the intern: 'I guess now we know why his kidneys weren't working.'

The urethra *was* blocked – by the prostate gland. The prostate surrounds the urethra, and when it enlarges, as it often does with age, it can impinge on the narrow outlet, obstructing and ultimately blocking it so that no urine can pass. As the trapped liquid filled the bladder, stretching it far beyond its normal capacity, the pressure shut down the patient's kidneys. Just hours after the obstruction was relieved, Charlie's potassium began to drop as the kidneys went back to work. Four hours later, his heart rate was up over sixty. By the next morning, the abdominal pain, probably caused by his hugely distended bladder, had eased. When he left the hospital three days later, his potassium and heart rate were normal and his kidneys nearly so. He would have to keep the tube in his bladder until the obstructed tube could be opened.

In the hours before his diagnosis, Charlie was seen by at least two nurses and three doctors. He had complained of abdominal pain. How is it possible that none of these doctors or nurses noticed that his bladder, normally the size of a hockey puck, was the size of a football? Charlie's a slender man, over six feet tall and weighing only 140 pounds. His belly is normally flat. I didn't see him that day, but I'm guessing it was distended and tender. No one noticed, I suspect, because no one looked.

No one examined Charlie Jackson – until it was almost too late.

## *The Delirious Doctor*

As a practising physician, I understand the temptation to skip the physical exam. A sick patient comes in and you are so focused on the thing that you are certain might kill him that you don't think of looking at anything else. There's a kind of anxiety, a controlled adrenaline-fed panic, when facing a patient who could die before your eyes. You pore over the labs and the studies. You get the consult. You send him to the ICU. But you don't examine him. That's not what doctors do anymore, in part because they no longer know how.

So thoroughly has this lesson been absorbed that doctors – those in training and out – often don't even notice when the loss of this creaky old antique makes a classic diagnosis impossible. I frequently attend medical conferences with the hope of finding cases for my newspaper column. I ran across a perfect example of this at a recent conference of the Society of General Internal Medicine, a gathering of academic physicians.

Judy Reemsma, a third-year resident, stood by her poster in the rabbit warren of partitions that make up the display halls where residents and medical students display research and case reports. She spoke with confidence about the case presented in her poster. She should – in this case she was both the doctor who made the diagnosis and the patient.

During her second year of medical school Reemsma became ill and was taken to the emergency room by her fiancé, David DiSilva. Assigned to the case was Dr Jack McFarland, an emergency medicine resident and close friend of Judy's.

McFarland, tall and slender with a slight stoop to his shoulders, greeted his friend from the doorway one spring evening in 2004. 'What are you guys doing here?' he asked. It was strange to see her there. And shocking for her to be out of the usual scrubs and white coat, dressed in the flimsy johnny coat that marks you as the patient.

As McFarland traded quick pleasantries with David, he tried to assess Judy's condition. She looks okay, he thought. Her heart was racing; the tracings on the heart monitor sped by at 150 beats per minute. Her blood pressure was high and though she did appear anxious, she didn't look particularly sick.

And then she began to speak. A wild river of words poured from her mouth. Random phrases, meaningless sentences, rapid incoherent paragraphs. There were snatches of sense scattered throughout the discourse but they were nearly drowned in the rushed torrent of speech. McFarland was stunned. He looked at the young man, who nodded. This was why they'd come.

Judy had been fine all day, David told him. He had the day off from work and they'd been together most of the afternoon. She had classes in the morning. Came home and studied. They'd gone to the gym and then made dinner together. Afterward, she'd gone upstairs to study. Maybe an hour later she'd complained of stomach pains. And the computer screen looked blurry, she told him. She decided to go back to the bedroom and lie down.

Another hour later he'd heard her fall – he rushed upstairs and found her on the floor crying uncontrollably. When she spoke, her words made no sense and it was clear to him that she was confused. That's when he started to get scared. Coming here she'd been so unsteady on her feet that he'd practically had to carry her to the car.

The patient was twenty-seven, athletic, and had no significant medical problems. She was taking an antidepressant, Paxil, and had been given another, Elavil, to help her sleep. But, David added, Judy didn't like the way the Elavil made her feel so she didn't take it anymore. She didn't smoke, drank only occasionally, never used illicit drugs. As McFarland and Judy's fiancé went through her history, the patient moved restlessly on the gurney. At times she would try to answer the questions, but her speech was jumbled – a word salad carrying little useful information. She seemed unaware that she wasn't making any sense.

'I need to examine you; is that okay?' McFarland asked the patient tentatively. She nodded her consent. The lights in the room had been turned off and when the doctor turned on the light, Judy cried out and covered her eyes. 'Oh yeah, the light's been bothering her since we got here. That's why we turned it off,' her fiancé told him. McFarland reluctantly dimmed the lights. She had no fever. Her mouth was dry and her skin was quite warm though not sweaty. The rest of her exam was normal. He tried to perform a thorough neurological exam but the patient was too confused to cooperate. An EKG showed no abnormalities beyond the rapid heart rate.

McFarland thought carefully about his friend, now his patient. For almost anyone with a change in mental status, illicit drugs had to be at the top of the list of possible causes, as unlikely as that seemed in this case. In addition, she had been prescribed a medicine – Elavil – that could cause many of these symptoms when taken in large doses. She had a history of depression, and her fiancé had been out of town frequently over the past several months. Was she suicidal? Could she have taken an overdose? That could cause the rapid heart rate and confusion. He knew that a high dose of Elavil causes blood pressure to rise initially, but that the real danger came later when it can drop precipitously. Her pressure was high, dangerously so. Maybe she was in the early stages of the reaction. On the other hand, it was hard for McFarland to believe that his friend had been that depressed. She'd seemed fine when he saw her last.

Perhaps she didn't have simple depression – maybe she was bipolar and her antidepressant had moved her from depression to mania. That could cause the pressured speech, but would it cause the very high blood pressure? And he knew her; wouldn't he know if she was bipolar?

Or could she have too much thyroid hormone? The thyroid is the flesh-and-blood version of a carburetor – working to regulate how hard the body's machinery works. Too little of this hormone and the body slows down. Too much and it speeds up. That could cause the tachycardia and the hypertension and sometimes pressured speech and confusion.

He questioned the patient's fiancé. Had she ever shown signs of mania? She had a history of insomnia, and sleeplessness was one sign of both mania and thyroid overload – was she up all night? No, until this evening she'd been fine, he insisted. She had been depressed, but that had all but disappeared after starting the Paxil – and that was months ago. Her sleeping was no worse than usual.

David paused. There was one other thing: after dinner he'd felt a little funny too. Not as sick as Judy, but his heart had been racing and he'd felt a little nauseated – though he felt fine now. They'd eaten some lettuce from their garden that night. Could their symptoms have something to do with that? Hearing this, the resident immediately thought of a patient he'd seen not long ago who had eaten pesticide-tainted vegetables from his garden. That patient had nearly died. But he'd been much sicker than this young woman. Moreover, his symptoms were the opposite of hers; his heart rate had been slow, his blood pressure nearly undetectable. He'd lapsed into a coma not long after arriving in the emergency room – they'd had to intubate that patient because his lungs had filled with water. Overall, a very different clinical picture.

Still uncertain, the doctor ordered some routine blood tests to look for the presence of an infection or an imbalance in her blood chemistry. He checked the thyroid gland. He also ordered a urine test to look for illegal drugs and Elavil, the medication she had been prescribed for sleep.

As the doctor waited for the results of the tests he'd ordered, the patient became more and more agitated. She kept getting out of bed and walking into the chaotic hub of the emergency room. Once she put on gloves and picked up the chart of another patient as if she were at work. Several times the nurses had to guide her back to her own bed. Lying on her gurney, she seemed to talk to people who weren't there, pointing to and batting at creatures no one else could see. At times she quieted down, mumbling words that her fiancé couldn't understand.

The results of the tests dribbled in but provided no additional clues. The thyroid hormone was okay. The drug screen was completely negative. There was no trace of Elavil. What was going on?

By dawn the patient's blood pressure had come down into the normal range, but her heart rate remained high. She was less confused. But she was still far from normal. Was this part of some underlying illness? She had an MRI of her brain to look for evidence of a stroke, a CT of her chest to look for tiny clots. Both scans were normal. After four days the patient had completely recovered and she was discharged, her diagnosis still unknown.

At home, Judy was troubled by her brief episode of madness. The unanswered questions were frustrating.

That afternoon she wandered out to the garden to do some weeding, and her attention was immediately drawn to an uninvited guest growing in her lettuce patch. Among the green and purple leaves she and her fiancé had planted were several strikingly beautiful white flowers, blossoms that hadn't been there before and that she was certain she'd never sown. Could the early tendrils of this plant have been mistaken for lettuce and ended up in her salad? She pulled the three plants up by their roots, put them in a Baggie, then drove to a nearby nursery.

As she pulled the plants from the bag to show the owner, the woman exclaimed, 'Don't touch those plants! They're highly toxic. That's jimsonweed.' Also known as the devil's trumpet and sometimes as loco weed, this plant has been known to cause a temporary kind of madness in man and beast for centuries, the woman explained. The symptoms caused by the active ingredient found in this plant are so well known that there is a mnemonic widely taught in medical school to identify its symptoms: mad as a hatter, blind as a bat, dry as a bone, red as a beet, hot as a hare.

As it turned out, the patient had had all the classic symptoms: the plant's toxin makes you blind as a bat because it makes the pupils dilate. (This chemical is still used by ophthalmologists for that very purpose.) And she was quite flushed, according to her fiancé. McFarland missed both symptoms in the emergency room because he had turned down the lights to alleviate his friend's discomfort. Her mouth and skin were noted to be dry and of course the madness was clear, but this wasn't enough to make a diagnosis. By the time the other doctors in the hospital saw her, most of these characteristic symptoms had resolved.

I asked Dr McFarland why he thought he had missed such a classic presentation of this well-described syndrome. 'I've thought about that. A lot, actually. I think my friendship with the patient made it difficult for me to really put on my doctor's hat. I never was quite able to see her as a patient.' The doctor-patient relationship requires a certain distance that the resident wasn't able to impose on his friend. 'You want to sort of keep your eyes averted, intellectually, when you're taking care of someone you know. You need to dig and yet it's uncomfortable.'

But there's something else going on here too. McFarland didn't insist on being able to turn on the light to examine his patient completely. Would he have been that blasé if the patient had refused to allow him to take her blood or urine for tests or balked at the idea of a CT scan? Why didn't he insist on having the light up so that he could perform the exam properly? Could it be that he didn't believe that the physical exam would provide any useful information that would allow him to make a diagnosis? Ultimately, of course, that loss of faith becomes a self-fulfilling prophecy. If you don't expect to see something, how hard are you likely to look?

And because he didn't insist on seeing her in the light, he didn't notice that she was flushed or that her eyes were oddly dilated in the bright light of the room. Opting to leave her in the dark, he unintentionally left himself there as well. He missed two essential clues that might have allowed him to solve the mystery of her illness.

## *The Science of the Senses*

It's been over fifteen years since Salvatore Mangione published his groundbreaking studies on the loss of physical exam skills among physicians. The studies have prompted active and passionate debate but little action, and even as these skills wither in the subsequent generations of doctors, we still have no idea what effect this change may have had on our ability to take care of our patients. Can technology replace these skills? Or will the loss of the exam damage our ability to make a timely diagnosis? With few studies done, we have no better idea of that now than we did in 1993. But anecdotal information suggests that there is a great deal being lost.

Doctors are not known for their rapid embrace of the new. Medicine has held on to the paper chart long after virtually every other industry and profession has made room for electronic efficiency. Physicians are so reluctant to change the way they practise medicine that it takes on average seventeen years for techniques well established by research – such as giving an aspirin to a patient having a heart attack – to be adopted by even half of those in practice. In other words, it usually requires an entire generation of doctors to turn over for a single new practice to become routine, part of medical 'tradition.'

Medical training itself has not effectively changed since the end of the nineteenth century, when Sir William Osler developed the hospital-based residency system as a method for standardising and institutionalising medical apprenticeship. Changes that have been imposed on doctors – for example, the eighty-hour working week – have been derided and abhorred by them from coast to coast.

And yet physicians and even patients have seemed willing, even eager, to abandon the physical exam, painstakingly developed over the past two centuries, and allow its erosion to advance unchecked. Undoubtedly medicine's characteristic conservatism has contributed to this loss. The almost pathological unwillingness to change the way new doctors are trained in the face of a rapidly transforming environment has helped bring about one of the most radical changes to how medicine is practised in its history.

Nonetheless, over these years there has also been a growing sense that the physical exam can make an important contribution to our ability to understand the patient and his disease. With this acceptance has come a new set of once unaskable questions: Which parts of the physical exam are valuable and worth saving? Which parts could and should be disposed of? And once we get a better handle on which are worth saving, how can we incorporate this into the education of our new doctors?

In the next few chapters I will examine each of the several parts of the physical exam, looking at the way each works to provide clues to the mystery of the diagnosis. We'll look at each part in the order we are taught to perform them: first by observation, then by touching, then by listening. Each method of evaluating the patient directly through our senses provides immediate and essential information. Each has its own limitations.

Once the exam is broken into its component parts, can we then identify which parts are important and useful and should be kept and what turns out not to be so valuable after all? If it is possible, if we can separate out those parts of the physical exam that are useful and discard the parts that are not, we will be left with a physical exam that is leaner but keener. If not, and the physical exam is lost, we will end up with a health care system that is slower, less effective, and more expensive – a high-tech, low-touch system that fails patients along with the doctors who care for them.

# CHAPTER FIVE

## *Seeing Is Believing*

Dr Stanley Wainapel walks carefully to the door of his office to greet his first patient of the day. It's a brutally humid July morning and even here, deep in the recesses of Montefiore Medical Center in the Bronx, a heavy dampness has overwhelmed the air-conditioning. Wainapel is a tall man in his early sixties. A shock of sumptuous white hair frames a handsome round face creased with lines that deepen when he smiles. His light brown eyes are magnified behind black wire-rimmed glasses that he adjusts frequently.

Wainapel runs the Department of Rehabilitation Medicine at Montefiore. He introduces himself to Anna Delano, the heavyset, middle-aged woman who has come to see him about her painful knees. As she makes her way to the chair in front of his desk and carefully lowers herself into it, he commends her for braving the humidity and apologises for the ineffective air-conditioning.

Anna looks up at Wainapel still standing in the doorway. 'Are you talking to me?' she asks, voicing her confusion in a nasal New York accent. 'Because, you know, you're not looking at me.'

Wainapel whips his head around to the spot where the voice now originates. Embarrassed, he smiles, revealing a deep dimple. 'Sorry,' he tells her, 'I have a vision problem.'

Here's the nature of Stanley Wainapel's vision problem: he's blind. Wainapel was born with a form of retinitis pigmentosa, a rare genetic disorder that started him out in life with severe night blindness and tunnel vision. Over the years, the narrow windows through which he could once see became progressively smaller until they finally closed completely, leaving him unable to perceive any colour or shape, and very little light. In his right eye, his 'good' eye, he can sometimes detect movement. In his left eye – nothing.

Because of the indolent course of this disease, Wainapel could see well enough to make it through college, medical school, a four-year residency in rehabilitation medicine, and the start of an extremely productive academic career. Wainapel says he feels certain that his visual defect hasn't kept him from being a good doctor. A successful career culminating in his current role as director of rehabilitation medicine, and a crowded schedule, suggest that he's right. My question is: how is that possible?

Vision has long been considered the most valuable of our five senses. Biologically it is certainly preeminent. More than 50 per cent of the human brain is devoted to sight. Thinking may be how Descartes knew his world, but for the rest of us, seeing is believing. We trust what our eyes tell us. When Chico Marx, pretending to be Rufus T. Firefly (Groucho Marx) in the 1933 film *Duck Soup,* is caught red-handed with another woman, he denies the obvious infidelity and demands indignantly, 'Who are you going to believe? Me or your own eyes?' It's funny because for most of us, that's no choice at all.

The same is true in medicine. William Osler emphasised the importance of observation in medicine: 'We miss more by not seeing than by not knowing,' he taught his students. Even the language of patient care emphasises the central role of vision. We 'see' patients in the office; we 'watch' them overnight in the hospital. We tell patients what to 'look out' for. We 'oversee' their care.

Well before a doctor begins the refined manoeuvres that make up what is usually considered the physical exam, she will start to collect information on the patient as soon as she lays eyes on him. Is he young or old? Does he look healthy or sick? How does he walk? Is he in pain?

Once the patient actually gets to the exam room, much of the physical exam relies on what doctors can see – they look at the skin and the eyes, peer into the ears and mouth. They check the colour of the tongue, the nails, the stools. Many of the tools used to perform the exam allow better views of the ears, nose, mouth; the equipment used to measure blood pressure, temperature, oxygen saturation, and blood glucose report this data visually. The tests ordered to provide additional information about the patient often convert that data into a visual form: diagnostic imaging is the most obvious, but an EKG is a visual representation of the electrical activity of the heart, and an electroencephalogram (EEG) represents the working circuitry of the brain. Of course, these studies are often interpreted by specialists – doctors don't always read them themselves. Still, given the importance of sight in medicine, it's hard to imagine making a diagnosis without it. How can a doctor 'see' patients if she can't see the patient?

No one seems to know how many blind physicians there are practising in the United States. A Google search ('blind physician') turns up a dozen names. Reading up on those that I found, I see that most work in specialties like psychiatry, where routine patient contact consists primarily of listening and talking. A couple, like Wainapel, have gone into rehabilitation medicine. I wanted to meet Stanley Wainapel to understand the value of vision in the practice of medicine and in making a diagnosis. Who would better know the true worth of that sense than one who once had the ability to see and now must work without it?

Confronted with the patient's confusion that morning, Wainapel deftly deflects the woman's question with humour. 'I'm not looking at you because you are so beautiful I had to turn my eyes away.' They both laugh and once the discomfort of the moment passes, Wainapel moves confidently back to his desk and starts asking the patient about her knee pain.

It started almost a year ago, she tells him, and has been getting steadily worse. She'd seen her doctor. He sent her to two surgeons. Predictably both recommended surgery. She came to Wainapel because that option had no appeal. 'I've lost thirty pounds and that helps, but not enough. It's hard for me to even walk. Now I've got to use a cane.' She also complains of wrist pain for the past couple of weeks. As she tells her story Wainapel sits forward slightly, his head cocked, eyes fastened on the patient's face – a picture of close attention. He asks her a few questions and jots down notes on a pad. From where I'm sitting I can't see what he's writing, but I notice that he uses his left thumb to keep his place on the page as he writes up the details of her complaint.

After reviewing her medical history and medications, he asks her to sit on the examining table that takes up the other half of the room. I watch as he skillfully touches and manoeuvres her shoulders, elbows, wrists, and hands to identify the source of her wrist pain. 'It really hurts right there,' she tells him as he holds her wrist. 'Here? Okay. Hmm.' Eyes closed, Wainapel lightly touches her forearm to identify the origin of the pain. 'That's between the ulnar stylus [the prominent bony bump on the pinky side of your wrist] and the pisiform [the furthest outside bone of the wrist]. Hmmm. No numbness? No weakness?' (No and no.) 'Sounds like a sprain of the ulnar collateral ligament. No trauma?' (No.)

Wainapel moves on to her knees. He assesses their range of motion – her gasps reveal how much her knees hurt with even ordinary movement. He feels her ankles and feet; he checks for evidence of swelling and joint instability. He asks her to lie on her back so that he can evaluate her hip joint. Sometimes pain felt in the knee actually originates in the hip. But not in this case. She has full range of painless motion in the hips.

They return to their seats and Wainapel walks her through his thought process. The wrist pain probably comes from a sprain – perhaps from the way she pushes herself up from a chair to stand. Still, it's essential to make certain it's not a fracture. One of the small bones in the wrist can break and pain may be the only clue – so she'll need an X-ray. About the knees – he'll need to get the X-ray report from her orthopaedic surgeon. Until then he recommends physiotherapy, Naprosyn (an anti-inflammatory medication like ibuprofen), and a trial of a glucosamine and chondroitin combination, an over-the-counter remedy sold in health food stores for joint pain. He reviews the evidence on this second medicine: 'Studies have shown that the glucosamine and chondroitin combination doesn't repair joints, but it can bring some relief in the group of patients with the most painful knees. If it can help you move, why not give it a try?'

As he speaks, I look around the office. I wasn't surprised that Anna hadn't immediately known that her doctor was blind. Nothing about him or his office suggests that he has any disability at all. In addition to the usual framed degrees, his office walls are covered with colourful oil paintings, photographs, posters. Bookshelves loaded with medical textbooks and references cover an entire wall. Wainapel wears glasses – not dark glasses, regular glasses. And his eyes appear to focus on your face when he's speaking – so long as you don't move. The only clue to his vision deficit is the two white canes discreetly tucked away against the bookshelf.

Before the patient leaves, Wainapel dictates a letter to the patient's doctor. He easily reels off a precise summary of all the patient has told him. 'This way they can see that I have no secrets and that I remember everything. That it's my eyes that are affected – not my brain. And, of course, if I make a mistake, the patient has a chance to correct me.' I steal a look at the notes he'd taken while talking to the patient. They are unreadable – not in the clichéd way that doctors' handwriting often is, with squiggles and lines that require careful deciphering. Despite his best efforts he's written his notes so that the dozens of lines of large loopy script are overlaid on top of one another – condensed into a single line of thick, indecipherable scribbles. I'm surprised. It's easy to forget that he can't see. Luckily he doesn't depend on these handwritten messes. He routinely dictates his notes on the visit and they are typed and placed in the chart. If for some reason the dictation system doesn't work and his report is lost, Wainapel tells me earnestly, his secretary can use these notes to reconstruct it. Hearing this, I make no comment. He can't see the mess his notes have become and it doesn't seem necessary for me to point it out to him.

As the visit draws to a close, Wainapel calls in his secretary, who writes out the prescriptions along with the referral for physiotherapy. She positions Wainapel's hand so that he can sign the pages in the proper spot, then walks the patient out to her desk to schedule a follow-up appointment. In all, a perfectly ordinary encounter.

After the patient has gone I ask Wainapel why he didn't let her know he was blind before she came – if only to prevent that awkward social moment. He seems surprised by the question. 'Why should I?' he asks. 'If it were important for my work as a physician I would, but it's not.' With a sly smile he adds, 'If I can find the patient, chances are excellent I'll be able to help them. With me, locating the patient is the hardest part of my job.

'Observation is certainly the most important component of the physical exam, but there are other ways to observe than with your eyes,' Wainapel tells me. He is a good listener, he points out. He prides himself on his ability to get the full history, to allow the patient to tell him what he has, and considers himself an expert in the physical examination of the musculoskeletal system.

'I knew I was going to be blind and so I made my choices based on that,' he adds. 'And because of these choices – my specialty for one – I think I can be an excellent doctor who happens to be blind. I don't know that I could say that if I were in another specialty.' What makes this specialty better? 'A lot of things. There's the obvious: the patients who come in to see me have problems in parts of their bodies that I can examine directly. I'd make a rotten surgeon or ophthalmologist – they need to be able to see in order to do their job. I can do mine with my hands, my ears, and most importantly, my brain.'

Comparing his practice to my own, I get a sense of how his specialty allows him to excel despite his limitations. The patients who come to see him are in pain, but the causes are chronic, not acute. The arm or leg they complain of is unlikely to be broken or infected or bleeding. He's not that kind of doctor. And because of the chronic nature of their problems, he has the time to accurately diagnose and treat most of the patients who come to see him.

And yet even within this specialty there are cases where the loss of sight has made a diagnosis difficult. Wainapel tells me about an elderly woman who'd come to him for rehabilitation after hip replacement surgery. Before her surgery she had been active and healthy, she reported – limited only by the pain in her hip. After the surgery she remained weak and unsteady on her feet despite weeks of rehab. She still needed a walker to keep from falling and had difficulty getting through the strengthening exercises. Wainapel was stumped. He examined her repeatedly. The surgical wound was well healed. The joint was freely mobile. Her strength and reflexes seemed normal, and yet she was unable to get around on her own.

A social worker provided the clue that solved the case for him. She was struck by the fixed, sad expression on the woman's face. Could she have Parkinson's disease? she asked Wainapel. It was a good thought – and something he couldn't have seen. 'I walked over to the patient and by golly she had cogwheeling and everything.' Cogwheeling is a jerky motion in the joint when that joint is passively moved – a cardinal symptom of Parkinson's. The disease slows voluntary movement and causes instability. No wonder she wasn't getting better. Once her newly diagnosed Parkinson's was treated, the patient improved rapidly.

Of course, from Wainapel's and the patient's point of view, this case wasn't a failure but a success. He was able to help this woman return to her previous state of vigour and activity – eventually. And yet the case shows that even in the narrow range of patients seen in this specialty clinic, there are those for whom sight plays an important and irreplaceable role. It was a success ultimately because in the patient population Wainapel cares for, there is time to figure things out. That is not always the case in other specialties.

## *The Look of Illness*

In medicine, sight becomes essential when rapid assessment and action are required. You can't imagine, for example, a blind emergency room physician. In an emergency you need to be able to collect information about the patient rapidly, efficiently. You never know what is going to come through the door and so you have to be ready for anything. Throughout medical school and residency training, I was told repeatedly that I needed to learn what 'sick' looks like because it would provide one of the most important clues about how ill a patient really was.

This is not a new idea. Some of the earliest writings we have are devoted to describing this look. Hippocrates begins his work on prognosis with this clue: 'If the patient's normal appearance is preserved, this is best; just as the more abnormal it is, the worse it is.' He goes on to describe the face of someone who is going to die: the nose is sharp, he tells us, the eyes sunken, the temples fallen in, the skin stretched and dry with a dusky colour. Hippocrates approaches the difficulty of caring for a patient who is too sick to survive with the same pragmatism that characterises the oath that still carries his name: 'By realising and announcing beforehand which patients were going to die, [the physician] would absolve himself from any blame.' This wisdom has been handed down through the centuries of medicine in all its various forms.

By the time most doctors finish training, they have at least one story about the patients who taught them what sick looks like. It's one of those rites of passage that can't be forgotten. Jennifer Henderson was the patient who taught me the look of the critically ill. And it was in caring for her that I discovered the unexpected limitations of this assessment. Caring for Jennifer, I learned that recognising sick is only a first step.

I met her on my first night on call in my first year of training. I still remember the excitement and terror that long-anticipated event carried for me. Clark Atkins was the resident charged with supervising my training that first month. He had been an intern himself until just three days earlier when this new year had started and he had risen from intern to resident. Now it was Clark's turn to pass on what he'd learned. We hurried to see a new patient – Jennifer – who had already been moved out of the emergency room to a private room on the fourth floor.

One of the most important decisions that must be made about a patient, Clark instructed as we climbed the stairs to the patient's floor, is how much supervision and monitoring that patient will need. Emergency physicians are usually good at making this determination, but because it is so important, it's essential to see the patient for yourself to make certain you agree with their decision. I stopped to jot this down in the little book I kept for recording the secrets of patient care, then hurried to catch up.

Jennifer was sitting up in her bed, leaning forward, an arm planted on either side of her knees. A plastic oxygen mask fogged with breath arched over her nose and mouth like some modern version of a harem girl's veil. She looked up dully as we entered the room, distracted by the work it took to breathe. The thin chart from the ER said that she was thirty-one years old, but to my eyes she seemed much older.

She was a small woman – slender with delicate facial features coarsened by what had probably been a very hard life. Her curly bleached-blond hair was marred by a thick stripe of black at the part. Her eyes were a light blue colour that might have been strikingly beautiful once but now seemed washed out, lifeless. Her skin was tanned and leathery from the sun, and as she spoke a block of unexpected darkness along the line of her cigarette-stained teeth revealed that she had lost a couple. Her arms were wiry, her clavicles protruding, and the skin on her face looked a size too large. The muscles in her neck were prominent and contracted with every breath she took as she struggled to bring in enough air despite the oxygen provided by the mask.

Clark nodded at me encouragingly and I stepped up to the bed and introduced myself. I explained that we were her doctors while she was in the hospital and asked her why she had come. She hurt all over, she told me. She was a heroin addict. She was doing okay. Until last week. Then she got this headache. Her sentences came out in short bursts, a few words at a time, punctuated by deep breaths. She'd had night sweats. And a fever. And now she felt out of breath. All the time. And it hurt. When she had to breathe.

She suddenly looked up in distress, and her body convulsed in a paroxysm of coughing. She grabbed a tissue and held it against her mouth under the mask. She gasped for air as the spasm tore through her upper body. Tears streamed down her face. Finally she was quiet. She wiped her mouth with the tissue, then showed me the dark bloody sputum. 'I think. I'm dying,' she gasped, drying her face with the edge of the sheet. I tried to reassure her that she would be okay, but I worried that she might be right.

On exam, she had no fever but her heart was racing and she was breathing more rapidly than normal. And despite being on an oxygen mask getting

50 per cent oxygen (normal air contains only 20 per cent oxygen), she still wasn't getting enough. The oxygen saturation of her blood was 90 per cent (normal is 100 per cent). Her neck was stiff. She couldn't lower her chin to her chest, a sign suggestive of meningitis, an infection in the lining of the brain. When I listened to her chest, there were coarse crackling noises – like the noise of a crisp sheet of paper slowly being crumpled.

The blood work sent by the ER doctors showed an elevated white blood cell count. Her chest X-ray was dotted with white cloudlike masses a little smaller than golf balls.

At the nursing station Clark and I reviewed the data and tried to put the story together. She clearly had more than one infected organ system: she probably had pneumonia, and meningitis seemed likely too. As an intravenous drug user, Clark reminded me, she was at high risk for accidentally injecting bacteria from her skin directly into her bloodstream. From there, these aggressive bugs can go anywhere and infect almost any part of the body. It seemed likely that these bacteria had infected her lungs and possible that they had infected her heart and her brain as well. The emergency room docs had already started her on several broad-spectrum antibiotics. We needed to get a head CT and a lumbar puncture to look for an infection in her brain and an echo of her heart to look for infections there.

As I wrote the orders, Clark's pager went off. It was the ER. There was another admission waiting for us downstairs. He looked toward the patient's door, clearly torn about whether we were done thinking about this patient or not. When the beeper went off again he stood, reviewed what else had to be done, and left me to finish up as he ran down to the emergency room.

When I had finished my note, I put it in the chart and went in to see the patient once more. She was lying back on the bed now, but if anything she looked worse than she had earlier. Her hair was drenched with sweat and her chest heaved with every breath. I needed to go down to the ER but I couldn't bring myself to leave her alone. Did she really look worse or was it simply the anxiety of a brand-new intern? I didn't know, but what I did know was that I afraid to leave her room, afraid that she really was dying.

The respiratory therapist came in and gave the patient a breathing treatment with albuterol – a medicine to reduce wheezing. Desperate with uncertainty, I followed him out of the room and asked him how he thought she looked. 'I've seen worse,' he told me before hurrying off as his beeper sounded.

I stood frozen at the doorway. I didn't want to leave because she looked so sick and yet I couldn't think of anything to do. Why was I more worried than the resident or the respiratory therapist? They had certainly seen more sick people than I had. And yet I couldn't shake this concern. I pulled out the card on which I'd written Clark's pager number. I had to talk with him to figure out what I should do. Before I could dial the number, David Roer, the attending physician, strode up. He was in his early forties and had dark hair and an open pleasant face. He greeted me with his usual cheer and asked me about the patient. I gave him a brief report and told him of my concern and then followed him into the room. He spoke with Jennifer briefly, then did a quick physical exam. I trailed him back to the nursing station, eager to hear his assessment. 'This patient is on the verge of respiratory arrest.' His tone was kind, without a hint of reproach. 'She really needs to be in the ICU. She's probably going to need to be intubated.'

Hearing those words, shame flooded over me. And relief.

Of course this is what she needed. Why hadn't I thought of this? My cheeks burned as I buried myself in the business of transferring the patient to the intensive care unit. Once she'd been moved into her new home, I ran down to the emergency room to see our next admission. The rest of that call day was a blur of admitting more new patients, following up on studies, and getting sign-out on the patients cared for by the other house staff as they headed home.

By the time I had finished all the tasks on my to-do list and trudged up to the sixth-floor call room, the predawn sky was beginning to lighten. I was tired but couldn't sleep. I went through every step of what had happened with Jennifer and tried to figure out how I had gone so wrong in the plans I had so carefully put together for her – plans that didn't take into account her most pressing and life-threatening issue, her breathing. It was right in front of me. And when her condition worsened so quickly – as I think it must have – I saw too that she was sick, dangerously sick, the kind of sick I'd heard so much about. The real surprise to me was that recognising that she was sick had not helped me know what to do about it. I don't think I figured this out that night but what I learned over the course of that month – and relearned many times over the years of my specialty training – is that, as important as it is, recognising what 'sick' looks like is only the first step.

In fact, several studies have demonstrated that the recognition of what 'sick' is, while much touted by residents and many experienced physicians, has not been shown to be accurate or effective in guiding medical decision making. In one study done at Yale, John Mellors, then a fellow in infectious disease, followed 135 patients who came to the emergency room with a fever and no obvious source of infection. The decision that had to be made at that time was whether these patients had a virus – in which case they could safely be sent home for rest and TLC – or whether there was a chance that they had a bacterial infection that would require them to take antibiotics. All of the patients in the study had blood cultures and complete blood counts drawn, a chest X-ray and a urinalysis performed. The decision to either admit the patient or discharge him, with or without antibiotics, was made after all the results except for the blood cultures were reviewed.

All of the patients enrolled in the study were followed throughout the course of their illness. Then the researchers compared how sick the patients really were with how sick the physicians had thought they were when they were initially seen in the ER. The doctors were wrong far more often than they were right. Many of the patients who were judged to look quite ill and were admitted were discharged soon afterward with no medical interventions taken. And four patients, assessed as being 'not toxic' and sent home without antibiotics, were ultimately found to have significant bacterial infections and had to be called back to the emergency room to get antibiotics. One of those patients died not long after being discharged from the emergency room, well before the doctors even had the chance to call him back in.

Other studies have also found that our instincts, our intuitive responses, to a 'sick'-appearing patient are frequently wrong. Recognition that a patient appears sick is important, it turns out, but it's not sufficient. As the Mellor study showed, patients can look extremely sick and not have a dangerous illness. Other patients, and this is particularly true of the elderly, can look remarkably well – at least for a while – despite a life-threatening infection. How sick a patient looks is just a clue, a single piece of data. Alone it is practically meaningless.

So what will help predict sickness? Concrete measures. Abnormal vital signs are key – a blood pressure that is too low or too high, a heart rate or respiratory rate that is too fast or too slow. Abnormal skin colour or mental status. We are very good observers of abnormality. However, we often respond immediately and viscerally to a patient's condition before we've even identified the abnormality that's the cause of the concern. The fear I felt in Jennifer's room was such a response. I recognised sick but hadn't gone the next essential step of identifying what was causing the fear and so I didn't know what to treat.

When the attending first saw Jennifer, he immediately recognised that she was dangerously ill. He then noted the abnormal respiratory rate, the effort she was expending to breathe. She was using the muscles in her neck and shoulders to perform an act that is normally simple and effortless. Moreover, despite the hard work she was doing, she still was not getting enough oxygen into her bloodstream. These are ominous signals. As a medical student I had read about how patients working this hard to breathe can tire out and die. I knew it and yet that knowledge didn't help me. I saw – it's probably how I knew she was sick – but I didn't recognise what I saw and so was unable to figure out what to do.

I followed Jennifer's course over the next week. As predicted, she wasn't able to sustain the effort it took to breathe and was intubated the next morning. Her blood cultures grew *Staphylococcus aureus*, an aggressive and destructive bacterium that lives on the skin. It is a disastrously common infection among intravenous drug users. Despite the powerful antibiotics, she continued to deteriorate. Her blood pressure dropped so that she needed medications to keep her blood circulating effectively. Then her kidneys failed. Her blood stopped clotting. After seven days in the ICU, Jennifer's heart and lungs failed her and she died.

I don't think the delay in getting Jennifer to the ICU had a major impact on her prognosis. I made important mistakes in my training – we all do – mistakes that hasten or even cause death in those at the boundary between life and death. But I don't count Jennifer among my mistakes. She had a severe infection and precious little reserve. Nevertheless, I think of her often. Those minutes of terror and confusion I felt standing powerless in her room served as a visceral reminder throughout my training (and even now, occasionally) that the big picture isn't enough in medicine; that the overall impression of a patient is worthless without looking further and paying attention to the specific measurements of health or sickness that were behind the impression in the first place.

Research into human perception reveals that we have developed a remarkable ability to quickly gather visual data and come to a conclusion without even noticing the steps we take to get there. Studies in perception show that this rapid automatic use of our eyes is by far the most efficient way to collect visual data. And most of the time, that's good enough. Not so in medicine. Inexperienced doctors, like my intern self, need to learn to make themselves work backward from the conclusions they reach, attend to the details that got them there, and translate what they see into the language and numbers of medicine. Only then can we at least try to help the patient.

## *Noticing What You See*

Sherlock Holmes perhaps expressed most succinctly the lesson I learned. 'I have trained myself,' Holmes tells his amanuensis, Dr John Watson, 'to notice what I see.' It's an important distinction.

'You have been in Afghanistan, I perceive.' With these first words Holmes initiated the quirky relationship with the man who would become his closest friend and most devoted follower. Watson, in London recovering from war wounds sustained in Afghanistan, is shocked by the man's declaration. How could he possibly have known this? Had he been told? 'Nothing of the sort. I *knew* you came from Afghanistan.' He retraces his reasoning. Watson's military bearing suggested some time spent in the armed services, Holmes tells him. The deep tan indicated a recent return; his wasted physique, some kind of intestinal fever. And his injured arm pointed to a war zone.

Of course it is an easy enough trick to pull off in fiction. However, Arthur Conan Doyle based his most famous character on a Scottish surgeon named Joseph Bell, for whom he'd worked during his medical training. Like Holmes, Bell frequently wore a deerstalker cap, smoked a pipe, and was often observed using a magnifying glass. But the most important trait they shared was a keen eye for detail combined with remarkable deductive powers.

Stories about Bell sound like snippets straight out of a Holmes story. In a preface to one of his books, Doyle describes his debt to Bell in developing Holmes as a character and provides examples of Bell's Holmes-like abilities. Seeing one patient, a young man in street clothes, Bell immediately asks the man if he was recently discharged from the military. He was. Was he a noncommissioned officer in the Highland Division? He was. Stationed in Barbados? Yes, how did he know all this? Like Holmes, Bell delighted in revealing his observations to the patient, the medical students, and the doctors observing him. Doyle quotes Bell's response: '"You see gentlemen," he explain[ed], "the man was a respectful man but did not remove his hat. They do not in the army, but he would have learned civilian ways had he been long discharged. He has an air of authority and he is obviously Scottish. As to Barbados, his complaint is elephantiasis, which is West Indian and not British." To his audience of Watsons it all seemed quite miraculous until it was explained and then it was simple enough. It is no wonder that after the study of such a character I used and amplified his method when in later life I tried to build up a scientific detective.'

Doyle clearly recognised that Bell's powers of observation were extraordinary. He referred to himself and the other doctors who witnessed these remarkable instances of detection as 'Watsons.' Yet Holmes and his model, Bell, firmly believed that this kind of close observation of significant details could be taught and sought to instruct those around them. 'From close observation and deduction you can make a correct diagnosis of any and every case,' Bell wrote in a letter to his now famous student, Arthur Conan Doyle. With practice, he suggested, the power of observation can be sharpened, improved. Doctors, he seemed to suggest, can teach themselves to 'notice what they see.'

## *Learning How to See*

Medical schools across the US have recently joined ranks with the historic Joseph Bell in striving to teach medical students to be better observers. One of the first efforts came from Yale. Dr Irwin Braverman, a professor of dermatology for over fifty years, had long been frustrated by the difficulty students had in describing findings of the skin. It might have been a knowledge deficit – easily remedied with books, pictures, and tests. But Braverman suspected that what his students principally lacked was the skill of close observation. Too often they wanted to cut straight to the answer without paying attention to the details that took them there.

'You teach students to memorise lots of facts,' he told me. 'You say: "Look at this patient. Look at how he's standing. Look at his facial features. That particular pattern represents one disease, and this pattern represents another." We teach those patterns so that the next time the doctor comes across it, he or she comes up with a diagnosis.' What's missing, says Braverman, is how to think when an oddity appears. That requires careful and detailed observation. After years of teaching he still wasn't certain he'd found the best way to communicate that complex set of skills.

In 1998 Braverman came up with a way to teach this skill. What if he taught these young medical students how to observe in a context where they wouldn't need any specialised knowledge and so could focus on skills that couldn't be learned from a book, where the teaching would force students to focus on process, not content? He realised that he had a perfect classroom right in his own backyard, in Yale's Center for British Art. The course, now part of the curriculum, requires first-year medical students to hone their powers of observation on paintings rather than patients.

As I entered the cool soft light of the museum's atrium, a dozen first-year students were standing around in small groups, waiting to enter the conference room to find out what they were doing in this unusual setting. Braverman, a round-faced man with a comb-over and an impish smile, sat at the head of a long table of lustrous dark wood like a folksy CEO of some big corporation. Their job that afternoon, he told them, was to look at the pictures they were assigned to and then just describe them. Not too hard, right? He looked around hopefully. A few students sitting near him smiled and nodded enthusiastically. The rest of the table was a harder sell. 'It's always like that,' Braverman told me as we followed the students up the stairs to the third floor, where most of the nineteenth-century paintings he liked to use were on exhibit. 'A handful of students either get it right away or are just habitually enthusiastic. The rest of the students here need to be convinced. But, you watch, by the end of the afternoon, I'll have some converts. Wait and see.'

Once stationed at their assigned pictures the students reviewed the rest of the rules. They were not to read the little labels next to the paintings. They'd have ten minutes to look at the pictures and then together the class would discuss the images, one by one. Each of the pictures would have a story to tell. It was the student's job to figure out what that story was and relate it to the rest of the group, using only concrete descriptive terms. If you think a character looks sad, he told them, figure out what you are seeing that makes you think that and describe it. If you think that the picture suggests a certain place or class, describe the details that lead you to that conclusion.

A tall young man with a sweet face and a prominent Adam's apple peered at the image of a slender man whose upper torso was hanging limply over the side of a bed, his right hand touching the floor. His eyes were closed. Was he asleep? asked Braverman.

'No,' he announced decisively to his fellow medical students gathered around the scene. 'He could be drunk – he has a bottle in his hand – but he's not asleep. I think he's dead.' 'How do you know that?' asked Braverman.

'His colouring – it's not right. He looks green,' he answered thoughtfully. 'And there's death all around him.' He described the sad scene. The young man lies in a small, unadorned garret apartment. An indifferent landscape of rooftops dark in the changing light of a setting sun is silhouetted outside the narrow dirt-encrusted windows. Petals of a dying rose ornament the windowsill, their colour grey in the fading light. Shreds of torn papers are strewn across the floor. 'I think he's taken his own life,' he concluded triumphantly.

'Excellent,' agreed Braverman. Linda Friedlaender, curator of education, spoke briefly about the painting (*The Death of Chatterton*, Henry Wallis's rendition of the suicide of the seventeen-year-old poet of the eighteenth century, Thomas Chatterton) and then they moved on to the next painting.

After the class, Braverman and I talked over coffee about his innovative teaching technique. 'Astute observational skills are usually acquired only after several years of being in medical practice,' Dr Braverman said. 'Suddenly, all of the accumulated experience leads doctors to see things they have not been taught before. They become terrific observers – eventually. With this course, I hope to jump-start these special diagnostic skills right from the beginning.' Even though they're looking at paintings, not patients, what they learn here can be applied to medicine, Braverman asserted.

He knows this because he tested it. For two years Braverman had participants write a description of what they saw in a dozen photographs picturing individuals with visible abnormalities. After the class they were given a different set of photographs with the same instructions. Tests were scored based on the description of specific aspects of the photographed abnormalities. Correctly identifying the disease or condition did not affect the score; identifying and describing the visual data was all that counted. Before-and-after test scores were compared, and students improved by an average of 56 per cent after spending this afternoon in the museum.

To ensure that this was not simply due to better test-taking skills the second time around, the same two-part test was given to a group of students before and after a lecture on physical examination. These students also improved – you don't get to medical school if you can't learn how to take a test – but not nearly as much.

Even before I heard about this study, I knew from personal experience that these skills could be taught. I was in my third or fourth year of medical school when I suddenly started seeing people with abnormalities everywhere. It was as if I had suddenly been transported into a world populated with the ill, the injured, the aberrant. Of course they were there all along – why hadn't I seen them? Certainly knowledge plays a role. When you learn a new word or name, it suddenly seems to be everywhere.

But it's more than that. We are trained from a very early age to avert our eyes from abnormalities. Children are fascinated by people whose appearance differs from what they've come to expect. And we teach them to ignore that interest. My daughter, Tarpley, once asked a cashier if she was a man or a woman. My husband flushed with shame for the discomfort it caused the homely, hirsute woman. He apologised but recognised that the damage was done. Afterward he explained to our daughter just how much that kind of comment must have hurt the woman. She doesn't ask those kinds of questions anymore. She's learned not to stare.

Medical school forces you to undo that training. You mustn't avert your eyes from abnormality. You need to seek it out. You need to figure it out. And it doesn't just turn off when you leave your office. I frequently (quietly, I hope) point out to my husband pathology that I see on the street – the rolling gait of a man with an above-the-knee prosthesis; the strange grey-toned tan of a man with iron overload syndrome, known as haemochromatosis, the schizophrenic woman's restless lips and mouth, a long-term side effect of many antipsychotics. I now live in a world filled with abnormality. It's fascinating.

How is it possible to see something without noticing it? Dr Marvin Chun, a professor in the Visual Cognitive Neuroscience Lab at Yale, has devoted his career to trying to answer that question. When I visited him on a warm autumn afternoon, he invited me to view a video already quite famous in his field of vision and attention. On a monitor I saw six adults standing in the midst of some strange game, their actions frozen by technology. There appeared to be two teams – one dressed in white, one in black. Each team had a basketball. Strangely, they weren't on a court but in the corridor of an unidentified office building. Closed lift doors were clearly visible in the background.

My task, once the video started, was to watch the white team and keep track of how many times the ball was passed between players – keeping separate counts of when it was passed overhead and when it was bounced from person to person. The image started to move and I kept my eyes glued to the white team's basketball as it was passed silently among the moving mass of black and white bodies. I got up to six overhead passes and one bounce pass and I lost track. Determined not to give up, I kept going until the thirty-second video was complete.

Eleven overhead passes and two bounce passes? I ventured. I told Chun that I got a little confused in the middle. Despite that, I'd done a good job, he told me. I missed only one overhead pass. Then he asked, 'Did you see anything unusual in the video?' Other than the unusual setting for the game, no, I saw nothing at all out of the ordinary.

'Did you see a gorilla in the video?'

A gorilla? No, I had definitely not seen a gorilla.

'I'm going to show you the video again, and this time, no counting, just look at the game.' He restarted the video. The white and black teams sprang back into action. Eighteen seconds into the game – around the time I lost my concentration – I saw someone (a woman, I find out later) in a gorilla suit enter the hallway court on the right. She strolled casually to the middle of the frame, beat her chest like a cartoon gorilla from a children's TV show, then calmly exited out of the left side of the picture. Her on-camera business lasted eight seconds and I hadn't seen her at all.

If you had asked me if I thought that I could miss a gorilla – or even a woman in a gorilla suit – strolling through the picture, I would have agreed that it was impossible to overlook such an extraordinary event. And yet I did. So did more than half of those who were given the same task by Daniel J. Simons in his lab at the University of Illinois at Urbana-Champaign. How is that possible?

We have tremendous faith in our ability to see what is in front of our eyes. And yet the world provides us with millions of examples that this is not the case. How often have you been unsuccessful in looking for an object and recruited the help of someone who finds it immediately right in front of you? Or had the embarrassing encounter with a friend who confronts you angrily after you 'ignored' his wave the night before while scanning for an open seat in a crowded movie theatre? According to the Federal Highway Administration, there are over six million car accidents every year. In many of these crashes, drivers claim that they had looked where they were going and simply hadn't seen the object with which they collided – evidence that people are regularly capable of not seeing what's in front of their eyes, what Sherlock Holmes might have called seeing without noticing.

Researchers call this phenomenon 'inattention blindness' because we often fail to notice an object or event simply because we are preoccupied with an attentionally demanding task. Our surprise when experiencing this very common event derives from a fundamental misunderstanding of how the brain works. We think of our eyes like movie cameras, capturing all that is before us as we choose what to focus on at the moment. We might not be paying attention to everything, but we assume, first, that we will be able to recognise any important event that occurs and, second, that, if necessary, we can always rewind the movie and play it back in the theatre of the mind. What we missed the first go-round would be noticed when we remembered the event.

Of course, that's not how it works. When asked about the gorilla in the basketball game, I had no memory of the beast. I searched my memory but I didn't remember him because I didn't see him. My attention was directed elsewhere.

There are qualities that make an object more likely to be seen. Chun tells me that if a naked man or woman had walked into the frame instead of a gorilla, the chance that I would notice the unexpected image would be much higher. Or if the gorilla had been bloody, or if he had moved or acted like a gorilla, I would have been more likely to see him. That's because there are some fundamental images that the mind recognises as important.

So what's going on here? Clearly the information is travelling through the eyes to the retina. And a functional MRI – one that reveals which areas of the brain are working in any given task – shows that the neurological signalling is getting the information to the right part of the brain – so you're definitely seeing it. But before this image can enter your awareness, another part of the brain jumps in to try to decide if this information is worthy of attention. And that judgment all depends on what you're looking for.

As it turns out, most of the time we see what we want to see, what we expect to see. Our ability to see objects or events that are unexpected and dissimilar to those that we are looking for is extremely limited.

To go back to the experiment with the ballplayers and the gorilla, my task was to follow the white-clad players and keep track of how often they threw the ball. Most viewers given that task fail to notice the gorilla. In the same experiment, subjects instructed instead to follow the ballplayers wearing black did see the gorilla. Because the gorilla was also black, it was closer to what they were looking for and so the image was able to get past the brain's gatekeepers and be noticed.

What happens to the visual information that enters the brain but doesn't get the attention of the subject's consciousness? Is it stored there, waiting for a second chance, like a delicious detail in a rerun episode of *The Simpsons*? Most research suggests not. If the sight doesn't capture one's attention initially, it's gone forever.

Based on research like this, Chun and many other researchers in this area now believe that the expectations of the viewer are the primary shapers of what is seen, and that the unexpected will often be missed. We become better seers when we have better expectations. When you are given a specific task – follow the ball as it's passed between members of the white team – you can predict what the expectations might be, and that observers are unlikely to see the passing gorilla because it's not in their set of expectations.

What about in situations where you are looking but the task is more complex – the way it is in real life, or in the hospital taking care of patients? If their theory is true, what you see and what you don't see will be shaped by what your experiences have led you to expect. Perhaps Osler was mistaken when he said that more diagnoses were missed because of not seeing than not knowing. Perhaps not knowing is what caused not seeing. Certainly that played a role in the case of Michael Kowalski.

## Great Expectations

Michael Kowalski was not a man who was easily frightened. And he could count on one hand the number of times he'd cried as an adult. But when Dr Keith Stoppard entered the room, he heard muffled, ragged breathing and, as his eyes adjusted to the dim light, he could see the massive man lying huddled in the bed. As unlikely as it seemed, Michael Kowalski, a fifty-two-year-old former college boxer, ex-army man, father of a marine, and all-around tough guy, was crying like a baby.

His wife, Maureen, a redheaded Valkyrie, stood by her husband's bed. Her face was darkened with freckles and lined by fatigue as she tenderly placed a cool, wet cloth on her husband's forehead. His short salt-and-pepper hair and unwaxed handlebar moustache lay plastered to his skin, and his round face was flushed and gleaming with the combination of sweat and tears. 'Doc, I'm scared,' he said, his raspy voice nearly a whisper. 'Can't you tell me what's wrong?' The woman squeezed her husband's hand in silent reassurance.

Stoppard, a third-year resident, didn't know what to say. He was worried. Mr Kowalski had been in the hospital for three days and Stoppard was no closer to figuring out what was making him so sick than he had been on the day the patient had been admitted.

On that first day, this had seemed like a pretty straightforward admission: a middle-aged outdoorsman sent in by his regular doctor for what looked like Lyme meningitis. Stoppard had spoken to the patient's doctor earlier that day and the case had seemed easy enough – get the lumbar puncture to confirm the diagnosis, then start intravenous antibiotics and watch him get better. But since then nothing had gone as he'd expected, and now he wasn't sure what to think or what to expect.

It was nearly midnight when Stoppard saw Kowalski in the emergency room that first night. The patient told him he'd started feeling sick about a week earlier. At first, he'd figured it was just the flu. He'd felt tired, his body stiff and achy. 'I was like an old man – I could barely get around,' he told the doctor in his low growl. But after two or three days of feeling lousy, he developed a strange, patterned fever. 'You could set your watch by these fevers,' he explained. 'Around four every afternoon I'd get real cold. I'd be shivering like mad. I'd load on the blankets but nothing I could do warmed me up. Then suddenly I'd be hot as hell. Sweaty. It was crazy.' His fevers would get up to 103°–104° every night, his wife, a nurse, added. By four in the morning he'd wake up drenched in sweat and have to change pajamas. By dawn, the fevers would subside – only to have the whole pattern repeat itself that afternoon.

Besides the fever, he told the resident, his neck felt stiff and painful, his head pounded, and a cough had left his throat raw. The joints in his legs, arms, and hands felt tight and sore. It was hard to move, to even get out of bed. Finally he went to see his regular doctor, Dr Dennis Huebner. After hearing the story and examining him, Huebner figured it was probably just a virus but decided to send off blood to test for Lyme to be safe. He knew the patient was at risk for the disease. Kowalski was an avid outdoorsman, spending as many weekends as he could hunting and fishing just outside Old Lyme, Connecticut, where the disease was endemic.

The patient had pulled off many ticks over the years, he acknowledged. But, he added, none lately; he'd been too busy to get out to the woods the past few months. Still, Huebner considered Lyme one of the diseases you just don't want to miss. If you get it early you can blast it with a week of antibiotics and it's gone. Miss it and the patient may need months of care. Huebner told the patient it was probably just something going around and he should call if the fevers persisted. He'd let him know if the Lyme test came back positive.

That night the fever came, right on time, and the next day the patient called Huebner, who reluctantly started the patient on doxycycline. 'He told me it was probably a virus,' the patient reported, 'but I felt like I was sick enough to need antibiotics. And he was okay with that. I took the pills, but the fever just kept coming. After a couple of days, the doc says to me: "Look, you're not getting better. You gotta go to the hospital." '

The patient considered himself a pretty healthy guy. He'd done his time in the army 'in the last war' (Vietnam), and now drove a truck for a local company. He had high blood pressure and his cholesterol was 'worse than the doc says it should be,' but he took his medicines regularly and had felt well 'until this crap started up.' On exam, in the emergency room, he had a fever of 103° and his heart was beating rapidly. The muscles of his neck were painful to the touch, but he could move his head freely. Just below his jaw he had several enlarged, painful lymph nodes. The joints in his hands and his knees were markedly tender but not red or swollen. Blood work sent by the ER showed an elevated white blood cell count and mildly abnormal liver enzymes.

The fever, painful neck, and pounding headache certainly pointed toward meningitis – a serious, potentially fatal infection. And untreated Lyme disease can progress to the brain, causing meningitis. But it wasn't a perfect fit: as awful as this guy felt, he wasn't as sick as the patients Stoppard had seen in the past with meningitis. With a fever this high, those patients were often too sick to talk. Despite the high fever, this patient was at times irritable, at other times funny, but very much awake and alert. The liver abnormalities weren't typical either. Well, maybe it was a viral meningitis – its course is much less severe than its bacterial counterpart and sometimes could drive up liver enzymes. In any case, they'd need to do a lumbar puncture. That would tell them if this was a meningitis and, if so, what was causing it.

But when Stoppard recommended this procedure, the patient blew up. He was already sick, already in pain, and now these doctors he'd never met before wanted to stick a needle in his back? No way. He would have to talk to his doctor. The patient's wife tried to persuade him but he was adamant: no procedures until he cleared it with his doctor – period. Huebner's partner was on call that night – would he speak with him? The patient sat up in the ER gurney and glared fiercely at the young resident: he would speak to his doctor and no one else. Defeated, Stoppard added high-dose intravenous antibiotics to the doxycycline he was already taking and waited anxiously for the morning and the certainty of the lumbar puncture.

Stoppard reached the doctor the first thing the next morning and he immediately called the patient. He needed this procedure, Huebner told him. They had to know if this was meningitis. The patient agreed, reluctantly, and the uncomfortable test was done. The results came back almost immediately – they were normal. There was no evidence of an infection in his brain. The Lyme test sent by the doctor days before came back that morning as well – it was also normal. He didn't have meningitis; he didn't have Lyme disease. They were back at square one.

One technique doctors use to make a diagnosis is to group symptoms, physical exam findings, and lab data and identify which are the most important and use them to try to find a recognisable pattern. This patient had many symptoms, but which were most important? Stoppard felt that the fever was key – it was extremely high and had this very distinct pattern. He wasn't so sure about the rest of them. But the fever, in combination with the enlarged lymph nodes and the elevated white blood cell count, clearly pointed to an infection. So where was this infection? What had they missed? Kowalski was on two strong antibiotics – but were they the right ones? At this point the team had no way of knowing. All they could do was keep looking.

In the emergency room they had drawn blood to try to grow the infecting bacteria, but so far they had shown nothing. They would need to be repeated whenever the patient spiked a fever – the time when the infectious agent was most likely to be found. A chest X-ray also done in the ER was normal but Stoppard ordered a second one – Kowalski had a fever, an elevated white blood cell count, and a cough – sometimes pneumonia can take a while to show up on an X-ray. He ordered tests to look for an infection in the patient's kidneys, his liver, his gallbladder. They revealed nothing.

On the other hand, Kowalski seemed to be getting better: he still had fevers every night, but they were 100°–101° – much lower than they had been at home or in the ER. And during the day, when the medical team made their rounds, Kowalski looked tired but said he felt okay – no headache, no body aches. Whatever he had, Stoppard was relieved to see that it was responding to the antibiotics.

Or so he'd thought until this afternoon, when the patient's temperature spiked to 104°, and the doctor had found him weeping in the darkened room. 'Tell me I'm not going to die,' he pleaded with the young doctor. 'Please help me.' He covered his head with the sheet and his shoulders heaved like a child's.

In that darkened hospital room, confronted with the patient weeping beneath his sheets, his wife white-faced with worry, Stoppard was overwhelmed. What if he couldn't figure this out? The day before, Dr Huebner had suggested that they send the patient to the big university hospital thirty miles away, but the resident had disagreed. He thought they'd find the answer. But right at that moment he was worried he had been wrong. To see this tough guy reduced to tears seemed a reproach of his skills, of his doctoring, of his judgment in keeping him here at this small community hospital far from the fellows and subspecialists at Yale.

Stoppard, now a nephrology fellow at the University of Pennsylvania, remembers that moment well. 'I didn't think he was going to die. But I couldn't promise that. And I couldn't lie to him, of course. But I wanted him to know that we were working as hard as we could to figure it out. And I felt pretty sure we would.'

He outlined the plan that he'd worked out with the infectious disease specialist brought onto the case. Infection was still the most likely cause of the fever, he told the patient and his wife; they just had to find it. A CT scan of the abdomen and pelvis and an MRI of the brain would show if there were infections hidden there. An ultrasound of his heart would help them look for unusual infections in the valves – infections that can take weeks to grow in cultures. None of these infections is common, Stoppard explained, but neither was a fever that didn't respond to a week of antibiotics.

And, while infection is the most common cause of fever, he continued, it's not the only one, by any means. Blood clots can produce fevers; so can some cancers. Finally, diseases of the connective tissues of the body – the joints, blood vessels, and muscles – can cause fevers and body aches. They would run some specialised blood tests to look for these diseases. Something was sure to show up, Stoppard assured his patient.

The cool air of the hallway hit his face when Stoppard finally left the room, and he realised he was almost as sweaty as the patient. He wrote the orders he'd told them about and waited for something to turn up.

But nothing did. The tests were done over the next two days as the fever kept its nightly schedule. The scans of the brain and body were normal – no clots, no infections, no other enlarged lymph nodes. The ultrasound of the heart was unremarkable. The blood cultures remained negative. Tests of his liver, which had been abnormal on his admission, remained abnormal, but hadn't worsened. One test stood out: the sedimentation rate. This is a very old test, and looks at how fast red blood cells sink to the bottom of a tiny capillary tube – a reflection of the amount of inflammation in the body. In this patient it was dramatically elevated. However, the test isn't very specific about what is causing the inflammation – it's one of the reasons it's not used very often. It could be an infection but it could also be cancer or one of the diseases of the connective tissues. They hadn't found any evidence of a cancer, and blood tests for lupus and rheumatoid arthritis – the two most common connective tissue disorders – were normal.

The resident was uncertain what to do next. Huebner once again brought up the possibility of transferring the patient to Yale. Kowalski had been in the hospital for almost a week, and they were still in the dark. Stoppard discussed the case with his colleagues and older, wiser physicians. Most of the tests they suggested had already been done. Then he spoke with Dr Alfred Berger. A youthful man with a broad Irish face and easy laugh, he was new to the faculty but had already become a favourite with the residents. After Stoppard went through the complicated story, Berger asked only one question: 'Does the patient have a rash?' No, they hadn't seen one, Stoppard answered. But why had he asked that? Berger smiled. It's all about patterns, he told the resident. The triad of persistent fever, joint pain, and a rash is the classic presentation of adult onset Still's disease, an unusual and poorly understood disease of the connective tissue.

Still's was first described in children, and in paediatrics it's now known as systemic juvenile rheumatoid arthritis. Young adults are the usual targets. There is no way to test for it. It is a diagnosis of exclusion – in other words, before you diagnose Still's you have to rule out everything else it could be. 'If that's what it is, it's a great diagnosis. It's rare and it's cool,' he exclaimed enthusiastically. 'Plus you've definitely got to know this one for the boards' (the tests you have to take in order to get licensed), the young teacher added as an afterthought.

In Still's, a rash is usually seen on the trunk and arms and is often only visible when the patient is febrile. Neither the patient nor his wife had said anything about a rash. Stoppard's team was on call that night so they would be able to look for the rash once the fever appeared.

They got that chance in just a couple of hours. Late that afternoon Stoppard got a call from the medical student on the team. 'The rash, the rash – he's got it!' she shouted excitedly. The medical student had told the patient and his wife to be on the lookout for a rash that afternoon. When she'd come to the room to check on him, Kowalski gave her a big smile, then said gruffly, 'Hey, Doc, wanna see a nice ass?' He'd turned and dropped his trousers to show a rash across his backside.

The resident hurried to the room. The rash was made up of painless, slightly raised, irregular patches of an unusual shade of pink; in textbooks, it's often described as salmon-coloured. The patient was started on prednisone, the usual treatment for Still's, and his response was a near instantaneous confirmation of the diagnosis. When he was given the first dose of medicine, his fever was 102.7° and the rash glowed. One hour later, both had completely disappeared.

The next morning the patient was up and dressed when Stoppard brought the team on rounds. His hair was combed, his moustache was waxed, and the car keys were on the bedside table. He was, he told them as soon as they walked in the door, ready to go home now. The fatigue, the muscle pains, the sore throat were completely gone. They wanted to keep him one more day – just to be sure – but the patient wouldn't hear of it. 'Aren't you sick of me yet? Because I sure as hell am sick of you guys.' Reluctantly, he agreed to stay until late afternoon, when the fever usually started, and when it didn't arrive he and his wife went home.

Why wasn't the rash obvious until that evening? Was that the first day he'd had it? In reviewing the chart, I saw that the attending had noted a rash several days earlier. At the time Huebner had attributed it to a simple skin infection and no one else had commented on it. And when asked, none of the team members recalled seeing the rash at all. It was outside their set of expectations. They simply didn't see it. Knowing what to look for makes it far more likely that you will find it.

The patient took prednisone for six months. He followed up with a rheumatologist who was familiar with the disease. She warned him that the disease recurs. It's been a couple of years and the disease reappears occasionally. 'I like the house cold when I sleep – always have – but when I wake up and my pillow is sweaty I know the Still's is on the warpath,' says Kowalski. 'But I don't let it slow me down.' He takes a week's worth of prednisone and again the symptoms disappear as quickly and mysteriously as they had that first time. He has to take it easy for a day or two, but knowing the diagnosis, understanding the course of the disease and what to expect, allows him to tolerate the symptoms with equanimity. The fear, the not knowing that made the fever so intolerable in the hospital, is gone. What's left is just the discomfort. 'I never even heard of that disease before I got it,' Kowalski told me, then added: 'To tell you the truth, I don't think my doctor had either.'

It's a truism in medicine that difficult diagnoses are most likely to be made by the most or least experienced doctors. The most senior have a broad set of experiences that allows them to consider many different possibilities. Because they are open to a wide variety of observations, fewer pertinent findings are filtered out. What about the novice? They have no expectations and there is some evidence that this lack of preset experience-based biases allows them to look more carefully at the entire picture.

Dr Marvin Chun points to an experiment conducted by his lab a couple of years ago. Participants in the study were shown two pictures; they were identical except that in one picture, a single element had been changed. The participants were shown one of two pairs of pictures. In one set the object that had been changed was central to the image. This picture showed a large room in which three people dressed in laboratory clothing are standing before a background of complicated machinery. In the first image, two support arms located just behind the people are painted bright yellow; in the next image that's changed – they're orange.

The second couplet featured a group of hot air balloons in the shape of farm animals. Well above them at the top of the photograph hovers a large hot air balloon with a clown's face painted on the side. A large, vibrant pink dot is visible on the clown's cheek. In the distance a second hot air balloon can be seen. A bright pink scarf flutters from the surface of the balloon. In the second image of that series, the hot pink spot and scarf disappear.

The researchers' hypothesis was that most viewers would notice the changes in the image of the laboratory immediately because the object that changes colour was located right behind the people at the centre of the image. The change in the second set of pictures, they thought, would be harder to see because the change was peripheral. They were right. Subjects needed much more time to identify the peripheral change. Experience has taught us that important information in a photograph is principally found in the centre, so that's where we look first.

The researchers took the experiment one step further. What if the picture defied our usual expectations? Would that change how quickly we were able to find the difference in the two images? To answer this question researchers showed another group of subjects the same pictures with one difference: this time the pictures were upside-down. In this experiment the subjects would have no experience in the new inverted world, no experience-based biases, and in this setting, Chun hypothesised, the change that was peripheral to the action would be just as obvious to the viewer as the change that was central to the content of the picture. In fact, that was the case. With the upside-down pictures it took about the same amount of time to identify either of the changes.

So the novice has no expectations, the expert has many expectations. Both states facilitate close observation. Where does that leave doctors (like me) in the middle – after our neophyte days but while still on the road to expertise?

This is an area of great interest to Chun and other researchers, and it's a hot topic in error reduction research. 'I don't know that we've found much that's useful yet,' admits Chun. 'I think the most important thing we've learned is that the control of this is primarily in the brain of the viewer.' He believes that drivers – and even doctors – need to be taught to direct their attention more broadly. When we focus too narrowly we will certainly miss something. 'It's all there for the taking. We just have to learn how to see it.'

After a morning spent with patients at his office in Montefiore, Dr Stanley Wainapel settled back in his chair and loosened his tie. 'People often ask me how I'm able to do the physical exam without my vision. Actually, for me, that's the easiest part. If you go to hear a heart murmur, what's the first thing you do? You close your eyes. You don't want anything else to interfere with your ability to hear. And if you thought you just felt a liver edge you close your eyes to feel it.' I found myself nodding in agreement – once again forgetting he couldn't see me.

He paused, thoughtfully adjusting his glasses. One of the lenses suddenly burst out of the frame. I jumped up and offered to help him find it but before I'd even finished my sentence he'd found the lens and put lens and glasses into his breast pocket.

'My patients learn pretty quickly that I don't see much, but here's the strange part.' He leaned forward and looked me squarely in the face. 'They still bring in their MRIs; they still want me to look at their X-rays. Why do they do that? They know I can't see them.' I considered this paradox – why bring pictures to a man you know is blind? 'They don't want me to see them. They don't care about that,' Wainapel explains. 'They want me to help them see what's going on. They want me to help them understand. That's really my job. Same as any doctor.'

## CHAPTER SIX

# *The Healing Touch*

The healing power of touch has long been part of Western culture. The prophet Elisha was said to have brought the dead to life with a simple touch. Jesus laid his hands on a leper and he was cured. His disciples were also granted this power to heal. Christian saints often performed miracles of healing by touch. And since Western monarchies were granted their power by divine right, many kings claimed this power as well. Until the eighteenth century, a single touch from the monarchs of England, Germany, or France was thought to be able to cure scrofula, a chronic infection of the skin – a therapy just as effective but far less painful than other available cures.

The use of the touch in medical diagnosis has had a spottier history. Hippocrates relished and eagerly employed the data provided by the senses. He wrote, 'It is the business of the physician to know in the first place things . . . most important, most easily known, which are to be perceived by the sight, touch, hearing, the nose and the tongue.' And texture, temperature, and contour were often provided in the description of patients and their diseases in his works. That approach to medicine was followed only intermittently until the Renaissance, and not until the Enlightenment was it fully re-embraced by physicians who sought to use the concrete data provided by the body to make medicine a true science in an age of scientific achievement. Ultimately it is the same quest for the precision and accuracy of a true science that has practically destroyed the physical exam. The doctor's touch seems primitive and uncertain when compared to what we can find out through the marvels of technology.

That's the perception, but is it true? There's mounting evidence that the hand of the doctor provides information that can't be gained from the cool eye cast by its technological replacements. Take, for example, the issue of screening for breast cancer. What can an exam pick up that can't be discerned by a machine? The machines in question – mammography, ultrasound, magnetic imaging – play a powerful role in the detection of breast cancers. But so does touch. Most breast cancers – well over 70 per cent – are detected by women who feel a lump in their breast. Mammograms account for another 20 per cent – clearly an important tool in the detection of this common disease. Yet studies suggest that the breast exams performed by a physician account for another 5 per cent of breast cancers detected – given the number of breast cancers in the US, that comes out to ten thousand cancers picked up on exam every year, making touch a surprisingly powerful tool as well.

The assessment of abdominal pain – one of the most common and problematic emergency room complaints – is another example where the physical exam may work better than even the best technology. Every year over three million patients come to an ER somewhere in the US complaining of pain in the belly. A quarter million of those patients end up in an operating room, having their appendix taken out. Most of the time, it's a good call – the surgeon will remove a diseased organ. But on average 20 per cent of those who take that trip to the OR will have what the surgeons call a negative appendix – that is, an appendix that is completely normal. For women the rate of unnecessary appendectomies can be twice that, up to 45 per cent in some studies. And these statistics have been unchanged for decades.

For many years this was considered an acceptable rate. Overall it was clear that early intervention was the safest way to deal with this potentially fatal disease and that the benefit of rushing patients with suspected appendicitis to the OR outweighed the potential harm of the unnecessary surgery.

Twenty years ago, Alfredo Alvarado, an emergency room physician in Florida, developed a method of separating patients who may have appendicitis from those whose pain probably comes from something else. Using the Alvarado score, doctors can then identify those with likely appendicitis, who can be taken directly to the operating room, from those with possible appendicitis, who should be watched. Alvarado considered three components in developing his score: three symptoms – nausea, anorexia, and abdominal pain that migrates to the right lower quadrant; three physical exam findings – fever, tenderness over the right lower quadrant, and the presence of what is known as 'rebound tenderness,' where the sudden release of pressure on the abdomen is more painful than the pressure itself; and a single test showing the number and type of white cells in the blood. Each factor present carries one or two points for a maximum score of 10. Those patients with scores of 7 or more probably have appendicitis and can go straight to the operating room. No further testing is needed. Those with scores of 4 and under probably don't have appendicitis at all and should be evaluated for other causes of abdominal pain. In studies, this system has been shown to reduce the rate of unnecessary appendectomies to less than 5 per cent.

The score is useful for those at the extremes – patients with a score of 4 or less and those with a score of 7 or more. But what do you do with those in the middle? Those who have signs and symptoms that suggest appendicitis but whose scores don't put them in the definite category? That's when technology comes in handy. CT scans can correctly distinguish between those who need surgery and those who don't almost 100 per cent of the time. Using both the Alvarado score and CT scans in cases when the diagnosis is unclear has been shown to be very effective and reduces the rate of negative appendectomies to nearly 1 per cent.

If the CT scan is so good at showing who needs to go to surgery, why not use it all the time? Why not take all patients with possible appendicitis pain straight to the CT scanner? In fact, that is what has happened. CT scans are routinely used to evaluate virtually all patients with abdominal pain. But a recent study suggests that this may not be the best strategy. Herbert Chen and others at the University of Wisconsin looked at the records of 411 patients diagnosed with appendicitis. Two thirds had a CT scan before going to the OR. In the other third, the decision to take the patient to surgery was made based on the history, physical exam, and laboratory findings. What they found was that those who had the CT scan had a much higher rate of complications than those who went straight to the operating room. And the rate of perforations was twice as high in those who had the test. Why? The authors speculate that it was the slower time to the OR. The third that didn't have the CT scan went to the OR within the first five hours of their arrival in the ER, while those who got the CT scan had to wait almost twice as long for surgery.

Despite the research, this is still a remarkably hard sell. In my community hospital Dr Jeff Sedlack is in charge of teaching general surgery residents. For years he lamented the fact that his trainees took virtually all patients with abdominal pain straight to the CT scanner, skipping the exam completely. He got tired of complaining, so eventually he decided to try something new. He set up a competition. Surgical residents would get one point for every patient with a suspected appendicitis that they examined and calculated an Alvarado score for. Patients who had a CT scan before being seen by the resident were disqualified. The trainee with the most points would win a small prize.

The residents took the competition seriously. One second-year resident got a bonus when he was able to persuade the ER doctor that the patient had a clear diagnosis of appendicitis without the expensive test. Instead of taking the patient to the CT scanner, the surgeons took him to the OR, where a pus-filled appendix was quickly removed. The competition was a tremendous success. The rate of CT scans went down, complications went down, and as an additional benefit, says Sedlack, the residents' exam skills improved dramatically. The next year the competition was brought back – by popular demand.

The presence of abdominal pain and tenderness can be extremely useful in making a diagnosis. Sometimes the inverse is also true: seeing a person in pain when that pain cannot be elicited by touching can also be informative.

## *The Dog That Didn't Bark*

It was July 1, my first month admitting patients as a resident. As an intern I had a resident at my side supervising my every move. Now I was the night float resident – taking admissions after the on-call team had reached their quota of daily admissions. My presence allowed them at least the possibility of sleep. It was exhilarating and a little scary being on my own this way. I knew there was always someone with more experience around – should I need them. Still, I was nervous.

I got my first call from the ER at around two a.m. A woman had been brought in from a nursing home by ambulance. Over the clatter of a busy ER the doctor spoke in the coarse, cryptic patois of medicine.

'We got a sick, demented eighty-seven-year-old female, name of Carlotta Davis. Sent from an ECF [extended care facility, or nursing home] with acute change in mental status. She's got a history of hypertension, CAD [heart disease], and a three-vessel CABG [heart bypass surgery] twenty years ago. She was out of it when they came to tuck her in, so they sent her over. Nothing on exam except a borderline low BP [blood pressure]. Labs showed a white count of sixteen [that's high] and a dirty [infected] urine. We sent her up on IV Cipro [an antibiotic] and a litre of saline [for the low blood pressure]. We're getting slammed down here so I gotta go.' And the line was quiet again.

Here's rule one you learn as a house officer – never just accept the emergency room's diagnosis. It's their job to determine if a patient is sick enough to be hospitalised. They like to give a patient a diagnosis because the paperwork asks for one and they often have a good idea of the problem. But they don't have the time or the resources to determine the diagnosis for any but the most obvious cases. Rule two: if the ER says someone is sick, go see them right away. They know what sick is.

'Mrs Davis,' I said softly as I entered the darkened room. I heard a long, deep moan. I turned up the light to reveal a tiny woman engulfed by pillows and blankets, moving restlessly in the bed. I crossed the room to the bedside and introduced myself. The patient lay with her eyes squeezed shut, shifting from side to side as if looking for a comfortable position. Her legs whispered to me as they slid back and forth continuously across the rough white sheets.

'Carlotta?' I tried again. No answer. I touched the thin skin of her wrists. She was warm and her pulse rapid but barely palpable. Her blood pressure was low – same as it had been in the ER.

'Can you open your eyes, Mrs Davis?' Again she didn't answer. If anything, she squeezed her lids even tighter, as if opening them would make the unbearable even more so. Just the moan and the near constant motion on the bed. Was it pain that was causing this movement, or delirium? Could be either. I listened to her heart, then slid my stethoscope beneath her bony back to listen to her lungs. I pulled back the covers and hospital gown to reveal an unexpectedly rounded stomach. Why wasn't her stomach as thin and as flat as the rest of her tiny slender frame? I rested my stethoscope on the mound lightly. Silence. I forced myself to listen for a full minute. Normally the gut is always making noise. Not this gut.

The emergency room doctors found that she had a urinary tract infection and were concerned it had spread to her blood. They'd already started her on intravenous antibiotics. It was a common cause for hospitalisation among the elderly and frail. And sometimes a severe illness can cause the gastrointestinal tract to shut down temporarily. Was that what was going on or was there something else? I carefully examined her abdomen. I felt no masses – no tenderness, either. She never flinched, never reacted to even the deepest touch. She was clearly in pain, but what I was doing didn't seem to affect it.

I pressed my fingers firmly down over her bladder. She had an infection here – was this the source of her pain? No reaction. I squeezed and thumped her flanks where the kidneys were hidden. Was the infection there too? No change in her restless movement. I finished my exam, carefully looking for any other potential sources of discomfort. No bedsores; no swollen, painful joints; no redness anywhere. Nothing that would account for the terrible restlessness and the haunting moans that escaped her lips every few minutes.

I had cared for many patients with urosepsis but none of them had looked like this. I ordered a small dose of morphine. We're taught not to treat pain until we know where it's coming from but I wanted to see if it gave her relief – if her distress was from some unfound source of pain. The morphine stopped the restless movement, but the moaning continued. I still didn't know whether it was pain or delirium.

According to the aides at the nursing home, the patient had complained of abdominal pain earlier that day. She may have had an infection in her urine but she didn't seem to have any pain in her bladder or kidney. What else could it be? In this age group cancer was likely. Did she have a colon cancer that was obstructing her bowels? Her stomach was soft, easy to examine, and I hadn't felt any of the firm linear masses that suggest stool trapped in the colon. A gallstone could cause fever and an elevated white blood cell count, but I would expect that to cause pain when I palpated her right side. There was none. Same with appendicitis, kidney stones, pancreatitis, a perforated viscus – all caused tremendous pain but those pains were usually localisable. I couldn't think of anything that could cause a pain this intense that couldn't be made worse with pressing.

And her blood pressure was still too low. I ordered another litre of normal saline. In the very ill, inadequate fluid in the bloodstream due to not eating or drinking or to excessive sweating can cause blood pressure to drop. Replacing that fluid will often restore a normal blood pressure. If her blood pressure didn't come up with this fluid, she'd have to go to the intensive care unit to get medicines to restore it.

I sought out Dr Cynthia Brown, the third-year resident on call in the ICU. Cynthia was a lively, down-to-earth redhead who had been a physiotherapist before going to medical school. Older than most of the residents, and like me, a southerner, she and I had bonded almost instantly. I found her at the nurses' station drinking hot tea and reviewing charts. She hadn't been to bed yet but looked remarkably awake and cheerful. She greeted me enthusiastically. I briefly laid out the case, running through my differential and my misgivings.

'There's something more going on but I can't figure out what. And I'm not even sure where to start. Do I send her down for a CT scan? And what of? If I don't get her blood pressure up, she's coming to you anyway.'

Cynthia thought for a moment.

'She has heart disease?' she asked.

She did.

'And her blood pressure is low?'

It was.

'Would you say that you thought she had pain that was out of proportion to what you found on physical exam?' she asked.

Absolutely.

'Those are the classic symptoms of ischaemic colitis.'

Like so many terms in medicine, the words themselves tell you much of what you need to know about this disease: ischaemia – from the Greek *isch,* restricted – and *haema,* meaning blood. Restricted blood flow to the colon. It's a disease most commonly seen in the elderly, often under conditions of a significant infection. I knew about this entity, of course. It's in my *Harrison's* – the textbook I used to learn about diseases. But the 'pain out of proportion to the exam' isn't in *Harrison's.* Or in any of the other textbooks I'd reviewed. It's part of the oral tradition in medicine, picked up – like so much else – the hard way, by not knowing. Still, I should have at least included it in my list of possibilities. She was a perfect setup for it. My face burned as I realised that, of course, ischaemic colitis was the most likely diagnosis. And I had missed it.

'Just remember, the reason you took this miserable, low-paying job was because of the education.' Cynthia smiled as she repeated back to me the words I'd said to her once as an intern. As I hurried back to the patient's room I boiled with frustration. How was I ever going to master all this? I read the textbooks, the little books of clinical pearls, the countless journal articles, and yet with a classic presentation of a classic little old lady disease I'd missed the boat. Internal medicine seemed suddenly, once again, completely overwhelming. It is vast; it is constantly changing; it is unmasterable. A resident I'd known during my intern year had recently shared with me her decision to leave internal medicine and go into dermatology. Why? I'd asked. She said, 'Because I want to be right more often.'

With the diagnosis of ischaemic colitis in mind, it was easy to reconstruct what must have happened. The patient had an infection, which caused her blood pressure to drop. She had hardened and narrowed arteries – that was why she had the heart bypass surgery years ago. Low blood pressure and bad arteries together can cause some parts of the body to be starved of new blood and oxygen. The pain she felt was the tissue dying for lack of oxygen. It's a terrible disease and often requires surgery. Mortality is high – in part because only those with multiple illnesses and poor overall health tend to develop this disease.

The room was quiet when I returned. The morphine finally allowed the patient to sleep or at least to stop moaning. And her blood pressure had crept up with the additional fluid. An X-ray confirmed the diagnosis of ischaemic colitis. I called the patient's primary doctor and, at his request, the surgeons.

Additional admissions sent me scrambling down to the emergency room. I returned a couple of hours later to see how the patient was doing and what the attending physician had done. She'd been evaluated by the surgical resident, who wanted to take her to the OR. New labs suggested that there was dead tissue that needed to be removed.

Her family did not agree to the operation. She had already made her wishes known to them – no extraordinary measures, no surgery. They would control her pain, the family instructed, and see what happened. If she survived, so be it; if not, at least let her slip away peacefully. Her daughter would be in as soon as she could get there. I went in to see the patient before I left that morning. The room was quiet but now filled with light from what looked to be a glorious summer day beyond the window. She lay unmoving on the bed; her eyes remained closed but the muscles of her face were finally relaxed. The delicate pale skin of her face draped gracefully over her cheekbones, like a sleeping beauty never found by her prince.

Although there was nothing I could do for her, I dropped by to see Carlotta the next night, and the night after that. She never woke up when I called her name or touched her thin shoulder. The room slowly filled up with cards, colourful drawings, and flowers. 'We love you Grammy,' neatly outlined in black and roughly crayoned in primary colours, was taped to the wall across from her bed so that it would be the first thing she saw when and if she opened her eyes. Toys stored on the deep window ledge suggested at least one grandchild or great-grandchild was a regular.

When I came by the fourth night the room was empty. The cards and drawing were gone; the bed, crisply made, waited for its next occupant. Standing in that doorway, I said my own goodbyes to this woman. This is how every doctor learns, often by standing at the bedside of the patients she didn't save. And this is how doctors pay their own private respects. I have diagnosed this disease and others similar to it, and every time I make the right call, I see Carlotta's face once more.

## *Hand to Hand, Mind to Mind*

Part of the romance, the appeal of the physical exam – at least for me – comes from the way it's taught. I learned from the individual physicians who instructed me. They, in turn, had learned it from the physicians who taught them, creating a line of transmission that extends backward, like genealogy, to the originator. Emphasising the personal nature of this transmission, the examination manoeuvres or techniques often carry the name of the doctor or sometimes nurse who created them. Spurling's sign, named for an early-twentieth-century American neurosurgeon, describes the manoeuvre Roy Glenwood Spurling developed to see if a pain in the arm or hand originated in the cervical spine. In this manoeuvre the head is tilted toward the side with the pain and then the physician presses straight down, compressing the soft discs between the bony vertebrae. If this reproduces the pain, reported Spurling in a paper published in 1944, the pain can be attributed to a pinched nerve in the neck, a useful tool in the days before MRI and still routinely taught as a way to evaluate arm pain.

Tinel's sign was named after a French neurologist, Jules Tinel. He developed the test while caring for World War I soldiers with injuries due to gunshot wounds. Frequently, once the wounds were healed, sensation and strength would still be limited due to damaged nerves to the region. Tinel would tap on the nerve just before it entered the injured extremity. If the patient felt tingling in the damaged area, said Tinel, the nerve was recovering and the soldier could expect to get back some sensation and use. These days, it's commonly taught as a method for diagnosing carpal tunnel syndrome, an overuse injury of the median nerve that causes numbness or tingling in the thumb, first finger, or second finger. If a tap on the wrist reproduces these symptoms, the patient is said to have carpal tunnel syndrome.

Here's the problem. Many of these manoeuvres don't work. Spurling's sign is no more predictive of cervical disc disease than flipping a coin. Many people will have pain with this kind of manoeuvre, but the pain could have many causes: rheumatoid arthritis, osteoarthritis, bone metastases from cancer. And many with a pinched nerve in the neck will have no pain. Still, it keeps getting taught.

Tinel's sign is just as worthless in diagnosing carpal tunnel syndrome. People who have carpal tunnel syndrome may have tingling when the nerve is tapped, but so will people with other problems. And many people with carpal tunnel syndrome won't feel the diagnostic tingling when tapped. So it can't reliably identify either those who have it or rule out those who don't.

The individual components of the physical examination were developed when physicians had few other means of diagnosing problems. Any sign or symptom that was found useful at the time was welcomed into the fold. Unlike modern (and expensive) high-tech tests or medications, there was no requirement for any of these exam techniques to be evaluated. And often, when these techniques were developed, there was no way to tell if the tests were right or not except by surgery or autopsy. As technology improved, so did our ability to test our tests. But we're only beginning to do that. In the meantime, doctors keep teaching them.

A colleague, Dr Tom Duffy, told me about a test I'd never heard of, and about a patient for whom it made an important difference. Michael Crosby was a young man – healthy and active with no medical problems at all. Michael remembered clearly the moment he became aware that he was ill. It was his second day of teaching. A new job, a new school. He was giving the class a quiz and as the students worked he paced between their desks.

Their heads were down, pens in hand, eyes moving from the words on the board to their own papers as they worked their way through the first test of the year.

He was a substitute teacher. And that morning he felt strangely nervous. He could feel his heart pounding in his chest and hear himself breathing in short, deep gasps. He'd trained for five years to get here; done internships in some of the worst neighbourhoods in urban upstate New York, and yet this middle-class ninth-grade Spanish class in rural Connecticut had him scared? His racing heart told him it was true.

But was this fear? All he knew was that it was hard to breathe. Really hard. And suddenly he was terrified. Breathing – the easiest, most natural thing in the world – all at once felt neither easy nor natural. He could feel himself go through the motion of breathing and yet the breath didn't seem to make it to his lungs. He felt sweat beading coolly on his face. His tie felt too tight around his neck. He glanced at the clock. Could he make it to the end of the period? He sat behind the desk at the front of the room and tried to relax.

The bell finally rang. The students dropped their papers onto his desk and clotted at the door. Crosby was right behind them.

The hallway to the school nurse's office seemed to stretch out into the distance. Every step was an effort. 'I can't breathe,' he croaked, once he finally made it to the tiny medical office. 'I feel sick.' Pat Howard, the school nurse, led him to a bed. He could hear her asking him questions, trying to get more information, but it was hard to speak. He felt like he was drowning on dry land. She removed his tie, then placed a mask over his mouth and nose. The cool rush of oxygen brought some relief. He remembered being loaded into an ambulance. When he opened his eyes again, he was in the emergency room surrounded by unknown faces.

He was quickly diagnosed with a massive pulmonary embolus. A blood clot from somewhere in his body had broken free and been carried through the circulation into the heart, then lodged in his lungs. He was started on blood thinners and admitted to the ICU where he could be monitored closely. As soon as he was stable the doctors turned their attention to the clot itself: where had it come from and why did he have it? They needed to know because another assault like that could kill him.

Clotting is something our lives depend on. But like so much in the body, context is everything. In the right place, at the right time, a blood clot can save your life by preventing uncontrolled bleeding. In another setting, that same clot can kill. Clots normally form at the site of any injury to a blood vessel. They can also form when blood stops moving; that's why anything that causes prolonged immobility, like travelling or being stuck in bed, increases the risk of a pathological clot. Pregnancy increases your risk. So do certain drugs and hormones. Some people have a genetic abnormality that makes their blood coagulate too readily. Finding the cause of a clot is crucial to estimating the risk of another.

So, his doctors looked: He had no clot in his legs – the most common source of abnormal blood clots. CT scans of his chest, abdomen, and pelvis likewise showed nothing. He hadn't travelled recently, hadn't been sick. He took no medicines and didn't smoke. His doctors sent off studies of his blood to look for any evidence that his blood coagulated too eagerly. Normal. They could find no reason for this otherwise healthy young man to develop a clot. He was discharged from the hospital after two weeks and told that he would have to be on warfarin, a drug that prevents blood from clotting, for the rest of his life. Without it the risk that he would have another clot was just too high.

It's difficult to be a patient with an illness that can't be explained. What made that uncertainty even worse was the new certainty that accompanied it – that he would have to take a blood-thinning medicine forever. He was twenty-three years old, a jock with a sport for every season. The blood-thinning medicine would protect him from another pulmonary embolus but in return he would have to avoid anything that could cause bleeding – including the games he loved.

The patient searched for an alternative and found my friend Tom Duffy, a haematologist at Yale University with a reputation as a great diagnostician. He hoped that Duffy could figure out what caused this devastating pulmonary embolism and possibly get him off the warfarin.

Duffy is a slender, fit man in his sixties with round tortoiseshell glasses, a preference for bow ties, and a precise, studied manner of speaking. He listened to the patient's story and then asked for a few more details: What kind of physical activity had he been doing in the weeks before the clot? He was alternating three days of weightlifting with two days of swimming or running. Had he taken any performance-enhancing drugs? The young man admitted that he had when he was younger but he'd taken nothing for years.

As he listened to the patient, Duffy considered the possibilities. The first set of doctors had done the usual testing, so this was going to be one of the unusual causes of pulmonary embolus. The scans done when he was in the hospital hadn't shown a clot in the vessels of his legs or trunk. A rare blood disease called paroxysmal nocturnal haemoglobinuria can cause blood clots in the liver, the spleen, or beneath the skin. The CT scan wouldn't have shown that. Could he have this rarity? Or could he have a myxoma, a rare type of tumour that grows in heart muscle, which can cause a clot within the heart itself? The physical exam might give some clues if these diseases were involved.

As the patient undressed for the exam, Duffy was struck by the highly developed muscles of his upper body. 'He looked like one of those young men in a fitness magazine,' he told me later. 'It was quite striking.' Otherwise his exam was completely normal: there were no extra sounds in his heart suggesting a tumour or anything else obstructing the flow of blood. His abdominal exam revealed no tenderness or enlargement that would suggest a clot hidden there.

Duffy looked at the patient again. He remembered something he'd learned in medical school many years before. He lifted the patient's arm until it was parallel to the floor. Carefully placing a finger over the pulse at the young man's wrist, he moved the arm so that it was pointed just slightly behind the patient. Then he asked the patient to tilt his head up, turn his face away from the elevated arm, and take a deep breath. When he did that, the pulse disappeared. When the patient looked forward again, the pulse returned. He repeated the manoeuvre. Again, the pulse disappeared when the patient turned his head and took a breath. Immediately Duffy suspected what had caused the clot.

The vessels that carry the blood from the heart to and from the shoulders and arms have to travel underneath the clavicle and above the top of the rib cage – through a very narrow space. The presence of an extra rib or hypertrophied muscles of the shoulder or neck can make this tight opening even tighter. This problem, known as thoracic outlet syndrome, is most commonly seen in young athletes who use their upper extremities extensively – baseball pitchers or weightlifters – or in workers who use their arms above the level of their shoulders – painters, wallpaper hangers, or teachers who write on a blackboard. For those with this condition, when the arm is elevated, the extra bone or muscle narrows the space between the two structures and the vessels that travel through them can be blocked. This patient was both a weightlifter and a teacher. He was a perfect setup.

Duffy set about to confirm his diagnosis and rule out any other cause of the clot. The blood work ruled out paroxysmal nocturnal haemoglobinuria. He got an MRI of the heart, which showed no tumour. An MRI taken while the patient lay with his arms above his head and his head turned away – the manoeuvre he'd done for Dr Duffy – showed that one of the large veins carrying blood from the arms back to the heart was partially obstructed. Duffy was right. He referred the patient to a surgeon who had experience with this unusual and difficult surgery and the patient had his first rib removed from each side the following summer. The next winter he was able to stop taking the warfarin. That was four years ago. He's been symptom-free ever since.

The value of any test or exam resides in its ability to reliably predict the presence or absence of disease. Many doctors wrote to me, after I published this story, to question the accuracy of the test Tom Duffy had performed, a manoeuvre known as Adson's test. I searched the published literature, and these doctors were right – there was nothing on it. It simply hadn't been studied. In other words, no one really knows how good the test is.

On the other hand, the test was fast, convenient. It was easy to perform and carried no risk. One of the doctors who wrote to me about the test offered the following perspective: 'Whether Adson's manoeuvre is accurate or not hardly matters. The fact is that Duffy thought of the diagnosis – and if the manoeuvre promotes that, then it's a good test.'

And yet if a particular exam is not reliable, how are doctors to judge the results they get? Can their findings be depended on? If the exam suggests the presence of a specific diagnosis, will it pan out? If, instead, it suggests the patient doesn't have the disease, can we rule it out?

We know how well many of the various technological tests work. For example, it's been shown that an ultrasound is less reliable than a CT scan. And doctors can take that into account when they consider the test results – especially if the findings they get don't support their own diagnostic hunches. But we don't have that kind of data on many of the tests that make up the physical exam. And even for those for which we do have objective testing, the findings are often not taught. The result is that when we perform the physical exam we have no idea how much faith to put into what we find. That uncertainty can lead to the wrong diagnosis. Far more often it leads doctors to ignore or omit the exam and its findings and skip directly to a test that the physician can feel more confident about.

'The real problem,' says Dr Steven McGee, who has collected and reviewed much of the research on the physical exam, 'is that there is all this tradition handed down to us and our poor medical students try to learn all of it. Then they find out that some part of it doesn't work and they throw the whole thing out. The truth is that there is a lot in the physical exam that turns out to be not terribly useful. But there are parts that are essential, even lifesaving.' McGee is part of a growing movement in research to assess the utility of various components of the physical exam.

The physical exam isn't perfect, McGee told me, and we are all very much aware of that these days. 'Our findings on physical exam feel like shades of grey while test results literally appear in black and white.' When we compare our own uncertainty with the confidence we feel when we look at a piece of paper – well, it's no wonder we prefer tests. 'But what you don't see on that piece of paper and what we often forget is that these tests in which we have placed our confidence aren't perfect either.' Take the chest X-ray. How reliable is that? One of the most basic findings we look for in a chest X-ray is the size of the heart – is it normal or is it large? A straightforward question and a chest X-ray should show that clearly enough. Having said that, if the same X-ray is read by more than one radiologist, how often will they agree about this simple finding?

Statisticians measure agreement using a tool called the kappa statistic. This takes into consideration the fact that sometimes with even random occurrences like flipping a coin, two people will agree or get the same answer merely by chance. To find real agreement rates you have to account for those that occur just by chance. So to use the example of two people flipping a coin, simple chance would have the coins both land on the same side about half the time. If the two coins were in agreement more often or less often, that would be their kappa statistic. You wouldn't expect any more than 50 per cent agreement and so the two coin tossers would be expected to have a kappa statistic of zero. On the other hand, if two individuals were looking at either a red card or a blue card and neither was colour-blind, you would expect them to agree virtually all the time. Their kappa statistic would approach 100.

So how do radiologists do when determining if a heart is a normal size or larger? Their kappa statistic is 48. In other words, once chance agreement is taken into consideration, there's a good chance the two radiologists will disagree at least some of the time. The same kind of disagreement occurs in other types of radiology – the problems with mammograms have been the most well described. Researchers calculated its kappa statistic as 47. Mammographers agreed with one another about 78 per cent of the time. Pathology is another area of notorious disagreements.

Even laboratory testing is far from perfect. *Clostridium difficile* is a bacterium that causes severe diarrhoea and requires treatment with antibiotics. Diagnosis is confirmed by detecting a toxin produced by the bacteria in the stool. When the test is positive, you can be certain that the patient has the disease. When the test is negative, however, it's far from clear that the patient doesn't have this infection. Studies show that up to one third of patients who have the infection will still have a negative test. Because it's an important diagnosis to make, routine practice in the hospital is to repeat the test up to three times. Only when all three tests are negative can you be certain that the patient doesn't have this potentially deadly infection.

What we've ended up with, says McGee, is a culture where test results have too much credibility and the good parts of the physical get too little. Neither is good for the patient. And we forget that for many diseases the diagnostic standard is still the physical examination: there is no test better than the physical exam to diagnose Parkinson's disease or Lou Gehrig's disease. Same with many dermatologic diseases. We need to weed out the useless components of the exam. Stop teaching those parts, says McGee. The rest can play an important role in diagnosis. We lose our skills, McGee suggests, at our patients' peril.

David Sackett, a Canadian physician considered the father of evidence-based medicine, has been one of the strongest advocates of a more evidence-based approach to the physical exam. In the 1990s he started working with the *Journal of the American Medical Association* to develop a series of articles called the Rational Clinical Exam. Each article in the series asks a question: does this patient have (some disease)? The article reviews the parts of the history and the exam and then provides the doctor with a measure of the test's accuracy and precision. The first article focused on ascites – fluid in the abdominal cavity. In the intervening years the series has looked at everything from asthma to appendicitis. It's been enormously successful, devotedly read and cited by physicians long frustrated by the vagaries of the physical exam.

For example, the gold-standard physical exam to find ascites, I was taught, was the puddle sign. In this exam, you ask the sick patient to get on his hands and knees, as if he were playing horsie with a child. Theoretically the free-flowing ascitic fluid in the abdomen would collect at the lowest part of the belly – the part hanging down. By striking that with your finger you would hear a dull sound if there was fluid there, a tympanic sound if there was only bowel there. It turns out that this embarrassing and uncomfortable test isn't very useful. What was shown to be a more effective test was to check for fluid when the patient was lying on his back. The patient puts his hand on the middle of his abdomen, holding the subcutaneous fat in place, and the doctor taps sharply on one side of the abdomen while feeling the other side. If there's fluid in the abdomen, you'll feel it slosh against the inner wall of the abdomen. If there's only abdominal fat, you will feel no movement.

I went to hear Steven McGee speak at a meeting of the American College of Physicians. The large room was filled to capacity. After the introduction, he walked up to the stage, a small man, trim and owlish, with horn-rimmed glasses hiding his eyes. He spoke in a quiet baritone about his own approach to making the physical exam worth doing again. Sometimes, the exam will give you all you need to make a diagnosis. Sometimes, he said, it will tell you what the patient doesn't have. You just have to know which parts you can depend on. 'Who uses Tinel's test when you're seeing a patient with hand numbness and tingling?' he asked the audience. Hands appeared across the room. Bad news, he told us. Not a good test. Asking the patient to show you where the symptoms occur on the hand is a better test. Those with carpal tunnel are most likely to point to the thumb and first two fingers. Finding decreased sensation on the thumb and first two fingers is a fast and simple technique that may help you make that diagnosis.

His goal, he told his audience, is to help doctors examine patients more confidently and accurately. 'Once versed in evidence-based physical diagnosis, clinicians can then settle many important questions at the time and place where they first arise – at the patient's bedside.'

When his talk was over I overheard snatches of conversation as the audience left the hall to go to their next lecture. There was excitement, hope, and passionate discussions of the accuracy and validity of favoured physical exam tests. As I walked through the double doors into the crowded hallway, I fell behind a group of young doctors and overheard their brief conversation on the talk. One tall, dark-haired young doctor nudged his friend with an elbow and said simply, 'As if.' Then laughed. I didn't see his face, but the meaning was clear: as if this research could change a fait accompli, the death of the physical exam. The others laughed with him. Another in the group said, 'Like I'm not going to get the test.' It was an abrupt reminder of the conservative nature of doctors. Changing this new status quo would be a challenge.

I thought again of my sister-in-law, Joanie, who'd offered to teach me on her own cancer. The gesture suggested she had far more confidence in the diagnostic potential of the physical exam than just about anyone in that lecture hall. Would she care if these skills were just allowed to die? Would she even notice? Can simply updating our armamentarium of physical exam techniques – eliminating those that don't work, buffing up those that do – be enough to reanimate the corpus of the physical exam? If not, what else might be needed?

## CHAPTER SEVEN

# *The Heart of the Matter*

I leaned forward in my seat and pressed the cheap plastic earpieces of the stethoscope deeper into my ears. I could hear the normal double knock of the heart at work, but there was another sound there too – one I didn't recognise. It was a quiet scratchy noise – regular, rhythmic, driving – like a percussionist thrumming out a rhythm on a washboard.

At the business end of the stethoscope I wore about my neck, the end that I would normally place on the patient's chest, the silver-dollar-sized disc was missing. In its place was a small black box made of cheap plastic, about the size of a pack of cigarettes. It was a lightweight radio receiver and the sounds I heard through the earpieces were being broadcast to me.

*What is that noise? I should know this.*

I sat among a dozen or so other doctors listening intently, trying to identify the cause of these abnormal sounds. All of us, medical school graduates, several years of specialty training and practice under our belts, were here at a class taught at the American College of Physicians conference, to relearn one of the fundamentals of the physical – the examination of the heart. I glanced at the woman next to me; her casually curly grey hair framed a brow wrinkled with concentration. She caught my look and smiled sheepishly. Clearly she too was stumped. A younger guy with oversized glasses stared intently at the floor.

'Who can tell me anything about what we're hearing?' asked Dr Vivian Obeso, the course leader. She scanned the faces of the dozen or so doctors who sat before her, on the other side of a life-sized mannequin of a young man. His chest was exposed, a sheet covered the rest of him, and his plastic legs were amputated mid-thigh. The missing end of our stethoscopes rested on the upper left side of the mannequin's chest, a couple of inches below the clavicle, demonstrating where the sound we heard would be coming from, had this plastic dummy been a living patient. The tiny class sat silent. Despite the age and years of experience of most of the doctors, there was an awkward pause as we hesitated to answer – it was a moment straight out of sixth grade. I knew from my own years of teaching medical residents that it's often hard to tell what that silence means. Is the question too hard? Or too easy? Both provoke the same uneasy hush. I still hadn't recognised the heart sound and suspected that was true of the others as well.

'All right. Don't tell me what you think it is – we'll get to that. Just describe the sound.' Obeso tried again. 'First, when does it occur? Is it systolic or diastolic?'

A normal heartbeat has two sounds separated by a very short period of what is usually silence – these two beats and the pause between them is known as systole (from the Greek word *systole,* which means contraction, so named by William Harvey when he first described the circular motion of the blood through the body in the seventeenth century). These are the noises made as the heart squeezes the blood into the lungs (the right side of the heart does this job) and into the general circulation (done by the left side of the heart). This double knock, onomatopoeia'd as lub-dup, is followed by another pause, this one often longer than the first. During the pause between lub-dups blood pours back into the heart, refilling each side before the next contraction. This longer pause is called diastole (from the Greek for drawing apart, because the heart enlarges as it relaxes and fills with blood). Because the activities in these two phases are so different, heart sounds are usually identified by where in this cycle they occur.

'Who can tell me? Systolic or diastolic?' The woman next to me looked up. 'It's both,' she offered quietly.

'Right. Did everybody hear that? There is both a systolic and a diastolic component.'

I listened again. Indeed, the staticky sound came between the lub and dup, but then reappeared in the time between beats.

The teacher continued: 'The patient is a young man who comes to the emergency room complaining of chest pain. This is his heart exam. Can you describe the sound?' A young man in the front row looked up. 'It's scratchy,' he said.

'Exactly right.' Obeso nodded. 'So what is this? There are three components to this sound. You don't always hear all three, but even just two of them will allow you to make this diagnosis.'

Three components? Oh right. I didn't recognise the sound but I do recognise the description. This must be pericarditis.

An inaudible voice from the front row spoke. 'Correct,' Dr Obeso said, flashing her remarkably white smile. 'This is pericarditis. What you are hearing is a pericardial rub – the result of an inflamed pericardium [the sac in which the heart sits] rubbing against the smooth muscle of the heart. Here's another patient with the same type of rub.' We listened again to a different recording, trying to store the noise somewhere in our individual brains so that we'd recognise it, if and when a patient with a heart that sounds like this walks into our offices one day.

The American College of Physicians started these refresher classes in clinical skills in 1995 with little more than a library of recommended titles and a couple of computer terminals. The current lab director, Dr Patrick Alguire, first began teaching at the lab a couple of years later when the college decided to add a course in performing skin biopsies and suturing – surgical procedures that many internists do infrequently enough to need a refresher. But, says Alguire, it soon became clear that doctors wanted help not just with these unusual procedures but with skills they need to use a whole lot more often.

First they added classes in the breast exam and genital exams using patient -instructors to teach these procedures on their own bodies – an innovation already commonplace in medical school. Over the next several years they added classes on how to examine different parts of the body: the muscles and joints, the eyes, the thyroid gland.

The diversifying syllabus, says Alguire, was a response to the growing evidence that physicians were entering practice with important gaps in their clinical skills, gaps that would be difficult to fill by simply reading. 'We saw from the very first course that there was a huge need for this kind of hands-on learning. When you finish your training you get out into practice and you are suddenly confronted with all the stuff maybe you didn't learn – or didn't learn well enough. It's the stuff you didn't know you didn't know – until you needed it. That's been the driving force behind the centre.' Perhaps not surprisingly, says Alguire, most of the students at the centre are young – doctors in their thirties and forties.

This is the first year the centre has offered the heart exam. Alguire had been looking for a way to include it for several years but hadn't found a good method to teach it. And then he saw Harvey – the electronic dummy I had spent my morning with. He thought it would be perfect for the doctors who had requested assistance with the heart exam. Seven classes were offered over the course of the conference that first year. All were filled; most had waiting lists. The word was that the course was worth the queue for the chance at an unoccupied seat, that it was an efficient and effective way to brush up on the basic cardiac exam skills.

The life-sized mannequin is capable of simulating a dozen different heart conditions, offering high-quality digital recordings of the sounds of the abnormal heart. It can show the pulses in the arteries of the neck and where, on the chest, the heart beats most forcefully. It reproduces the differences in the sound depending on where on the chest the microphone is placed. All these characteristics are essential clues for the clinical diagnosis of a wide variety of diseases of the heart. And unlike our catch-as-catch-can training in the hospital, this Harvey could teach them all – a kind of one-stop shopping for the heart exam.

Listening is the third and final sense we use routinely in the physical exam. Doctors often listen to the lungs and the gut. We strain to hear the first and last sounds of blood rushing through arteries narrowed by our blood pressure cuffs to look for hypertension. We listen to the vessels of the neck, searching for pathologic blockages in the arteries that carry blood from the heart to the brain, a potential source of strokes. We press our stethoscopes firmly into the belly beside and above the navel to check for sounds of turbulent flow into the kidneys – a cause of high blood pressure resistant to routine antihypertensive medication. But mostly, we use our stethoscopes to listen to the beating of the heart. Detecting deviations from the expected lub-dup is one of the oldest and most valuable tools we have for diagnosing important and sometimes life-threatening diseases of the heart.

In many ways, the heart exam stands as a symbol of the entire physical exam. It's not the most complicated exam – the neurological exam is the probably the most complex. Nor is it the most technically difficult exam – looking at the retina of the eye may get that honour. And it's not the most time-consuming exam – that would probably be the psychiatric exam. But the heart exam was the first examination developed in modern medicine and the one most strongly linked with the physician's role as diagnostician and caregiver.

Moreover, the heart exam is a subtle exercise and requires well-developed skills to detect the nuanced variations from expected heart sounds. A thorough understanding of the anatomy and physiology of the heart and the circulatory system is essential in interpreting these quiet deviations and identifying the lesion they suggest. As such, it has functioned as the proverbial canary in the coal mine, the first alert that physician skill and interest in the physical exam was waning.

Salvatore Mangione chose the heart exam to test in his 1992 study of doctors' skills not only because it was an area in which he had noted waning skills but also because of this position in the pantheon of examination abilities. He describes it as the 'tip of the iceberg' of the physical exam – the most apparent component to doctors and patients alike of this much larger practice, this sensual science of the body, the physical exam. Technology is eroding, melting away this ancient, massive, and essential part of the way a physician knows the human body.

If and when the physical exam is saved, says Mangione, we will know it when the heart exam is restored to its former preeminence as the signal of a highly skillful, well-trained physician.

## *A Different Way of Listening*

On my first day of medical school I was given the short white coat that signified my status as a student of medicine and my first stethoscope. These two symbols of my entrance into medicine were presented in very different manners. The white coat was given in a ceremony on a beautiful September morning in 1992. A sun-drenched hall was filled with rows of folding chairs for me, my ninety-nine new colleagues, and our families. The two deans of Yale Medical School, Gerald Burrow, head of the medical school, and Robert Gifford, the dean of students, stood at the front of the room, welcoming us into the profession. The late morning sun poured through a wall of windows, reflecting off the polished wood floors of the hall, suffusing the room with a fog of light. After a few words of welcome, Dean Gifford explained that the short white coat that we were about to receive indicated our status as medical students; these would be replaced in four years, upon graduation, by a full-length garment that signified our role as full physicians and teachers. Then each of us was called to the front of the hall to receive our own white coat. As we walked up the aisle, a brief bio was read, our first introduction to our peers for the next four years.

My husband squeezed my hand as my own name and credentials were read and I shuffled down the row of chairs to walk up the aisle, put on the crisp white jacket, and took my place among my new colleagues. Pride and excitement shone on everyone's face. When the last name was read, faculty and family joined in giving us a round of applause. It was a magnificent moment.

My first stethoscope had a far more ignominious entry into my life that day. After the ceremony we were sent off to finalise the complex logistics of registration. After filling out and signing a sheaf of forms, we were given our schedules and the key to our mailboxes. They were already overflowing with the typical packets of welcoming materials – sheets listing courses and books, still more forms to fill out for library and ID cards, manuals on policies and procedures, rule books, discount cards for local stores, and advertisements for various tools of the trade – and a stethoscope.

The stethoscope itself was one of those advertisements – a gift from Eli Lilly. If I received that gift today I would have different feelings about it, but this was before I had really thought much about the meaning of these gifts from the pharmaceutical industry. It came in a slender white box with the name of the manufacturer written in a tasteful script. It had the elegant proportions of a box from a jewellery store. I put everything down and picked up the box. Inside, the stethoscope lay draped on a black cardboard background contoured to keep the precious instrument in place.

Lifting the stethoscope out of the box, I was impressed by its heft. The disc at the end was polished chrome. The name of the drug company was written on the diaphragm – but that first day, I didn't even see it. Shiny grey rubber tubing extended from the disc and split, ending in a length of curved chrome and two grey rubber earpieces. Despite the elegant presentation, it was an ugly industrial object and yet I loved it. To me, it was far more important than the white jacket of the morning service. This was the real evidence of where I was going, the proof that at the end of all this there would be patients and healing – just as close to me as this disc was to these earpieces.

And yet as I think back on this I realise that this was my first clue to the status of the physical exam. The white coat, symbol of authority, knowledge, and progress, was the focus of the official welcome. The stethoscope, the symbol of the physical examination of the body, of our role as caregivers, was an industry-supported trinket – a freebie.

At home after that first day at school, I pulled the stethoscope out again. The silvery arms crossed and reached down as graceful as a dancer in first position. I put the soft rubber pieces to my ears, expecting them to sink into place. They didn't. I pulled the stethoscope off and looked at it once more. I tried again. Still awkward, still uncomfortable in my ears. I flipped the thing around so that the earpieces looked up at me like a leering cross-eyed sailor. I tried again. This time the earpieces fit snugly, the soft rubber adjusting to the contours of my ears so that all other noises were blocked.

I put the silver disc over my heart, head cocked, and I listened. I heard – nothing. I stood quietly. Still nothing. Was there something wrong with the stethoscope?

I took a deep breath. That I heard. I breathed again. The sound was clear, like the sound of wind passing through a hollow piece of plastic. Then I stood quietly listening, listening. After what seemed forever I felt, rather than heard, a quiet rhythmic pressure against my eardrums. I concentrated on that beat, then finally – somehow – was able to hear the now familiar lub-dup. This instrument required a different way of listening.

This was not going to be as easy as it looked.

On another morning, this one nearly two hundred years earlier in Paris, a young physician with the improbably delicate name of René-Théophile-Hyacinthe Laennec was confronted with the problem of examining a plump young woman with chest pain who was suspected of having a diseased heart. The year was 1816. The problem was one of logistics and propriety: how to evaluate the heart of this young woman. The recently developed practice of placing the ear directly on the chest of the patient seemed likely to be ineffective as well as improper. Other techniques of examination, also newly discovered – palpation (feeling the chest for the beat of the heart) and percussion (thumping the chest the way you might a melon) – were attempted but quite useless in this case, reported Laennec, 'on account of the great degree of fatness.'

'I recalled a well known acoustic phenomenon,' Laennec wrote several years later. 'If the ear is placed at one end of a log, the tap of a pin can be heard very distinctly at the other end. I imagined that this property of bodies could be applied to the case at hand. I took a paper notebook, made it into a tight roll, one end of which I applied to the precordial area [chest] and putting my ear to the other end, I was just as surprised as I was satisfied to hear the beating of the heart in a manner that was clearer and more distinct than I had ever heard it by the direct application of the ear.'

The utility of the device, ultimately called the stethoscope (from the

Greek *stethos*, chest), was immediately apparent to Laennec. It was the first technologic development enabling a 'view' into the inner workings of the living body. The device was so successful at transmitting the noises from inside the chest to his ear that Laennec devoted the rest of his career to better understanding the instrument and the body it revealed.

In Laennec's time, diseases were classified primarily on the basis of symptoms. An illness was defined by the subjective sensations experienced by patients. Doctors didn't examine patients; they interviewed them. What constituted a 'disease' then was assembled from a constellation of subjective symptoms and distinguished based on the type of symptoms, the sequence of their presentation, their severity and rhythm. Physical signs – derived from the pulse, touching, and observations of the skin and excreta – were contributory, but of much less importance.

At the turn of the nineteenth century, two new, closely linked ideas emerged that would change medicine forever. First was the growing understanding that disease was caused by the disruption of individual organ function. An Italian physician and teacher of anatomy, Giovanni Battista Morgagni, published a book called *On the Seats and Causes of Disease Investigated by Anatomy*, just a few years before Laennec was born. This revolutionary tome presented detailed drawings of diseased organs and then linked these abnormalities to clinical diseases. The connection between the diseased organs hidden within the body and clinically apparent diseases led to the second new idea: if diseases were caused by organ dysfunction, then they shouldn't be defined by their symptoms – too many diseases presented with the same kinds of symptoms. If the patient couldn't distinguish which organ was involved – and this was and remains true – then doctors had to find some way to identify the source of disease independent of the patient's story. For this they turned to the body itself, to the physical exam.

This new generation of physicians rejected the dependence on the vagaries of the patient's history. They argued that diseases could be classified based on changes that could be seen, felt, tasted, smelled, and heard by the doctor – changes that could be detected by an objective observation, independent of the patient's subjective account.

Laennec was a leader in this revolutionary reworking of the fundamental ideas of medicine. He used his new invention to find concrete, objective manifestations of disease. Others before him had developed some techniques that Laennec himself frequently made use of. But it was Laennec who made the greatest contribution to the radical new medicine, not only providing its first tool but making the link between what he was able to see and hear and the hidden dysfunction within the body.

Laennec was in the perfect place to do this too. He was the director of Necker Hospital, a small institution on the outskirts of Paris. Because of his position, he was able to follow hospital patients and their examination over the course of an entire hospital stay. All too often he could then correlate what he found on examination with what was revealed at autopsy. Laennec pioneered the way to link the pathologic changes caused by disease within the body to clinical information – the physical exam – collected from outside the body. His work put the physical exam at the forefront of the modern approach to medicine. Using his eyes, his ears, his stethoscope, the doctor became a detective – deducing the pathology within from observations made from without. Using the clues provided by symptoms described by the patient and the signs elicited and observed by the physician, the doctor-detective was able to track down the villain – the morbid processes within the body.

Laennec recorded each patient's physical exam in his daily notes, carefully tracking how the exam changed over time and incorporating these findings into the cases he reported. When the patient died – a common occurrence for those sick enough to go to the hospital – Laennec could identify the cause of the disease and the symptoms that revealed it. Once Laennec had made this link between the findings on physical exam and those at autopsy, he was able to diagnose similar patients in life with a precision rarely seen in previous centuries of physicians. Many of the diseases we now routinely identify by physical exam were first described by Laennec.

For example, Laennec was the first to diagnose emphysema. Others had seen the destructive nature of the disease on autopsy but Laennec linked the symptoms and physical findings to the pathological entity. The case involved a thirty-seven-year-old farmer who was admitted to the hospital in 1818 for worsening shortness of breath. Any exertion left him gasping for air. His hands, feet, and scrotum were hugely swollen and tinged with blue – cyanotic from a lack of oxygen. Laennec and his colleagues had seen these symptoms before. It was usually attributed to heart failure, where the heart becomes too weak to keep pumping out the quantity of blood sent back from the circulation, and fluid backs up – accumulating in lungs, abdomen, and extremities.

The patient's story of slowly worsening shortness of breath with any exertion combined with the clear evidence of this kind of overflow had the doctors at Necker convinced that the young farmer had heart failure. Laennec disagreed. He looked at the barrel chest of the farmer – and pictured the hyperinflated lungs he'd seen in cadavers with emphysema. He thumped the chest and heard it resonate – which suggested the lungs were filled with air – and yet noted that when he listened with his stethoscope, very little air could be heard moving in or out when the man breathed. Based on this, Laennec predicted that at autopsy the man would have a disease of the lungs, not of the heart.

They didn't have to wait long to find out. The farmer first came to Necker Hospital in May; he died just five months later – not of heart or lung disease but of smallpox. The autopsy showed, just as Laennec predicted, a normal heart. In the lungs, however, the delicate membranal lacework of the air exchange tissue had been ripped away, leaving large empty holes throughout – the now classic finding of emphysema.

One of the heart sounds first described and understood by Laennec was mitral stenosis – a pathological narrowing of one of the valves of the heart. He tells the story of a strapping young man, Louis Ponsard, sixteen years old, a gardener, who came to Laennec's hospital complaining of 'oppression and palpitations.' He was a short man, muscular, and according to Laennec, 'having all the appearance of splendid health.' Ponsard told the young doctor that two years earlier he 'was occupied in carrying some soil on a wheelbarrow. He was forcibly stopped in the midst of his work by a violent beating of the heart, accompanied by oppression, spitting of blood and nasal haemorrhage, coming on without any preceding discomfort.' The symptoms resolved later that day, Laennec writes, 'but they reappeared each time the patient attempted to take the slightest bit of exercise.'

When Laennec examined the patient, he noted a subtle vibration of the chest, what's called a thrill, between beats. This was accompanied by a murmur that Laennec describes as a 'sound [like that] produced by a file rubbing on wood.' Based on these signs and symptoms Laennec postulated that the young man suffered from 'ossification of the mitral valve,' what we now call mitral valve stenosis, or narrowing. When blood leaves the lungs, it passes through the mitral valve to enter the left ventricle on its way out into the body. In this disease, that passageway becomes narrowed and rigid. When there is a need for greater blood – during exertion – the normal valve is able to open wider to let the excess blood through. In this young man the valve was rigid, bonelike, and so couldn't expand to allow the greater quantity of blood through.

Understanding the problem this way allowed Laennec to treat the disease. If the problem was too much blood to make it through the narrowed valve, the available solution was to reduce the amount of blood. The young gardener was bled several times with a dramatic improvement in his symptoms.

This was probably one of the very few diseases for which the commonly applied treatment of bleeding may have been effective. Of course, the treatment is only temporary. The young gardener had to return to Necker several times over the next several years to be bled. And ultimately he had to change jobs. He became the servant of a priest, and with this reduced workload his symptoms became much more manageable. Laennec never heard from him again. Perhaps he lived happily ever after, but given what we know about mitral stenosis now, it's unlikely that he survived many years after his initial visits to Necker.

I learned about mitral stenosis the way I've learned so much of medicine – from my own mistakes. In fact, Laennec's discovery is what brought me to that makeshift classroom at the American College of Physicians conference. Like the dozen or so other doctors, I was there because I suddenly understood that despite years of training and practice, I still didn't know how to perform an adequate examination of the heart. Just like the doctors in the studies I'd read, I couldn't recognise some of the most basic abnormalities of the heart. I owe that discovery to Susan Sukhoo.

Susan was a slender woman of Indian extraction who'd been born and raised in Guyana, then immigrated to Miami with her husband some twenty years ago. She became my patient when she moved to Connecticut to live near her sisters after finding out that her husband was supporting a mistress. She was fifty-eight, had a little hypertension that was well controlled on a single medication, and many of our early visits focused on the consequences of her grief and depression.

Then she developed asthma.

She came to my office one frigid December morning looking her usual self – dressed simply but with a quiet elegance in tidy jeans, colourful T-shirt, and blazer. A single strand of pearls hugged the contours of her clavicles, showing off a youthful neck. Her hair was swept up in a simple knot at the back of her head, its smooth darkness only beginning to show traces of white. She smiled shyly at me as I entered the room and greeted her. 'I'm wheezeling,' she told me in the lilting inflections of her Guyanese-Indian accent. I wasn't exactly sure what she meant. 'When I walk, especially when it's cold out, I start wheezeling,' she explained, and then, like a caller on the radio show *Car Talk,* she began imitating the musical sound she heard when she breathed. She was wheezing.

The 'wheezeling' sometimes woke her up at night and she would have to sit up. A couple of nights she ended up sleeping in a chair because she felt like she couldn't breathe lying down. She had no chest pain, but sometimes felt chest tightness when she took a deep breath. These episodes lasted only a few minutes. After they resolved, she told me, she felt fine. She had recently had an upper respiratory tract infection and with further questioning thought the wheeze might have started when she was sick.

On exam, her blood pressure was normal. The amount of oxygen in her blood was fine. But there were diffuse wheezes throughout both lung fields. The breath came in with the normal whoosh of air flowing through a tube. But when exhaling, her chest was filled with a variety of musical sounds. This cacophony of different pitches and durations sounded like an orchestra of plastic horns warming up before a performance. Otherwise her exam was unremarkable.

Wheezing is caused by a transient constriction of the airways. Asthma is the most common cause of wheezing but it would be unusual for a woman this age with no history of this disease to suddenly develop it. Some infections can cause the airways of the lungs to overreact to sudden changes in air temperature or flow and that can cause wheezing – especially when you go from the warmth of a well-heated room into the frigid winter air or take a sudden deep breath. I gave her an inhaler to dampen the overreacting airways and assured her that it probably wouldn't last long. Wheezing and cough are common symptoms after a cold and usually resolve after a month or so. She'd had her cold several weeks earlier so I figured she must be on the tail end of the thing.

When I saw her next, a couple of months later, I asked her about the wheezing. Oh yes, she told me, 'the wheezeling was there every day.' She took a deep breath and let it out slowly. I could hear the wheeze from across the room. The inhaler was helpful, she added, and she used it almost every day. I wasn't sure what to make of this. We learn in medical school that 'all that wheezes isn't asthma,' but what then? Was this emphysema? She had never smoked, but her husband had. Could this be a so-called cardiac wheeze, where a weak heart can't pump all the blood that comes into it and so fluid gets backed up into the lungs, causing the wheeze? She hadn't had any chest pain, and her only risk factor for a heart attack (which can give you a weak heart) was her high blood pressure, and that had always been well controlled. She was from an area where TB was common – could this be an unusual manifestation of tuberculosis?

I got an EKG, which was normal. Reassured that she hadn't had a hidden heart attack, I also tested her for tuberculosis. In addition, I ordered some tests to do over the next couple of weeks to try to identify the cause of the wheeze. Pulmonary function tests would help distinguish asthma from emphysema or heart disease. All the other possible causes seemed far too unlikely in this extremely healthy woman. I also gave her another inhaler, this one containing steroids to reduce the frequency of what I still assumed was an atypical asthma.

She returned to the office a month later. 'Did anyone call to tell you I was in the hospital?' she asked. I'd heard nothing about it. It is a chronic problem in the community where I work. When a patient goes to the hospital – especially the other hospital in town – the doctor is often the last to know.

It happened in the middle of the night, she told me. She'd woken up drenched in sweat and gasping for air. A cough emerged from deep inside her chest. Her heart pounded so hard she felt the bed move with every beat. The chest tightness she'd felt when she'd first described her wheezing was back and much worse than it had ever been. She struggled to the phone – any exertion made her chest squeeze even tighter. She cried when she heard the siren, so grateful that help was close. In the ambulance and in the ER she'd been given albuterol, a medicine that relieves wheezing for patients with asthma. Normally it helped but that night it didn't seem to do anything.

An EKG showed she wasn't having a heart attack. A chest X-ray showed fluid in her lungs and they gave her a shot of something they told her was a medicine to help her pee out the extra fluid. Within an hour of getting that shot she started to feel better.

She stayed in the hospital for three days as her doctors tried to figure out why she had the fluid in her lungs. Dr Eric Holmboe, an internist on the teaching faculty, had diagnosed her on examination. His residents had called to tell him about the fifty-eight-year-old woman with poorly controlled, newly diagnosed asthma, and even before he saw her he was creating a list of diseases that could cause an asthma-like presentation. Whatever it was, he told me, he would have laid out money that it wasn't asthma.

When he listened to her heart, he heard the murmur Laennec had described. It was a quiet sound and could easily have been overlooked in a noisy emergency room. He could really only hear it when the patient lay on her left side so that the mitral valve was closer to the surface of the chest. And yet when he heard it, he knew she had mitral stenosis.

An ultrasound of her heart confirmed his diagnosis. The blood that would normally travel through that opening, to fill up the left ventricle – the main pumping chamber of the heart – couldn't get through the now tiny orifice. The opening, normally the size of a half dollar, had shrunk down so that it was smaller than a dime. The circulating blood couldn't all get through and so it backed up, flooding the lungs with fluid. 'The doctor in the hospital asked me if I had ever had rheumatic fever as a kid,' Susan told me, 'and I told him everybody in my family had it. But I hadn't thought about it for years and years.'

Rheumatic fever is an inflammatory complication of strep infection – often strep throat. Most often joints are the target. Days to weeks after an untreated case of strep throat, the patient will develop hot, swollen, and painful joints. It can be a single joint, multiple joints, or most strangely of all the inflammation can travel from one joint to another. The same inflammatory process can attack the heart as well. It is frequently undetected because it doesn't cause any symptoms – not for years.

In Susan's case the damage done as a child had slowly eaten away at her valve and by the time she developed 'asthma' the valve was nearly completely closed. She was scheduled to get a new mitral valve in a month, she told me that day.

Mitral stenosis – why hadn't I heard any evidence of this significant lesion during her heart exam? I placed my stethoscope on her chest, starting, as I had been taught, on the right, and worked my way to the left side of the sternum, then down to the middle of the rib cage, and then left again toward the edge. The lower left aspect of the chest is where this particular murmur is usually heard; it then travels to the far left side of the body. When I reached the lower left position I listened intently. I could barely hear – something. I had her lean forward, so that the heart would swing out, a little closer to the chest wall. There it was – a soft, low-pitched sound that came between heartbeats in diastole, rumbly and harsh and very, very quiet. I listened near the edge of the chest. I heard it there too. Now.

In my earlier exams I had completely missed this. I checked my previous notes – no mention of a murmur. It was a quiet sound and I hadn't done the kind of thorough exam I had been taught to do, so I hadn't heard it. I finished up my visit; I told her to let me know when she was to go into the hospital and I'd come visit her there.

Ultimately Susan's problem was resolved at the source. The tiny opening was widened. She had her scarred mitral valve removed and a metallic valve was inserted. Her heart was as good as new.

At home the night after I heard about Susan's diagnosis, and for many nights thereafter, I thought about this missed diagnosis. All those months of 'wheezeling' and shortness of breath and I'd been treating her as if she had asthma. She was getting worse right in front of my eyes as the aperture of the mitral valve approached a critical stage. It distressed me to know I could have figured it out too, if only I had done a proper exam. How many other diagnoses have I missed because of an inadequate examination of the heart? And I'm not alone. How many diagnoses have we all missed, because most of us don't have a clue about an adequate heart exam?

## *Putting the Ear to the Test*

But what if it's not our fault? If so few doctors can make this kind of diagnosis, maybe it's not possible. Just how good is the heart exam at picking up these defects anyway? As practised now, we know that it isn't very good at all. Few of the doctors in practice and in training are able to use the heart exam to make a correct diagnosis. We've come instead to depend on technology to make this diagnosis for us.

Echocardiography has been shown to be accurate in diagnosing many of the same diseases that the cardiac exam used to be good for. Small wonder then that the number of echocardiograms has increased so dramatically. The number of echos ordered almost doubled over a seven-year period – growing from 11 million a year in 1996 to 21 million a year in 2003. In one large multispecialty group in Boston the number of echos increased over 10 per cent over one year alone, with 9 per cent of all patients seen in the practice getting one. Is it simply that we no longer have any faith in our own ability to perform the exam, or is the exam fundamentally flawed and ready to be thrown over? Actually, studies show that the cardiac exam can be pretty darn good when done properly. In one study, five cardiologists were pitted against echocardiography in fifty-two patients with known valvular heart disease – one of the most difficult and important diagnoses we make when we examine the heart. The cardiologists had to correctly identify which of the four valves of the heart was affected and estimate the degree of damage. Each patient was also evaluated by echocardiography. How did the cardiologist do?

As in so many of these contests, the machine won. The echo was correct 95 to 100 per cent of the time. Yet the doctors put up a good fight. Their diagnoses were right between 70 and 90 per cent of the time. Other studies have shown similar results. That's certainly much better than the current crop of physicians if you believe Mangione's studies. The question is – is it good enough? Doctors and patients alike would probably say no. The ear and the stethoscope cannot replace the echo for locating the source of an abnormal heart sound when it's important.

But here's the thing: not all abnormal heart sounds are important. Up to 50 per cent of people who have a heart murmur – the most common abnormal heart sound – have completely normal hearts. These patients don't need additional testing. What we really need are doctors who are able to reliably distinguish between those who need more testing and those for whom further testing is simply a waste of their time and money. How well do we do here, where it really counts? Can we distinguish between those murmurs that need further evaluation and those that are benign or innocent? Cardiologists can. In a study done by Christine Attenhofer of the University Hospital in Zurich, cardiologists correctly identified ninety-eight out of one hundred pathologic heart sounds. Can primary care docs match that? Somewhat surprisingly, there's very little research done addressing this important question. One study done of emergency room physicians suggests that they can – though not as well as the subspecialists. In this study, two hundred patients with heart murmurs were evaluated by an ER physician. The physician took a history, examined the patient, and got a chest X-ray and an EKG. He then documented – in writing – whether the patient needed further evaluation or had an innocent murmur. After this evaluation all patients had echos. Of the two hundred patients, 65 per cent had normal echocardiograms and thus innocent murmurs. These ER doctors were able to identify those who didn't need additional studies nine times out of ten, erring mostly in sending too many patients with a normal heart for further evaluation. But they missed fourteen of the patients who had abnormal hearts.

Can we get better? Several studies have been done evaluating programmes designed to better teach the cardiac exam. Not surprisingly, all showed that if you teach these doctors-in-training, they will learn. One course used recorded sounds that participants were required to listen to five hundred times. Their test scores increased fourfold – from the downright pathetic 20 per cent correct to a respectable 85 per cent correct. Other studies had students examine actual patients who had a variety of heart murmurs. These doctors doubled their test scores. So it is a skill that can be learned. We have the tools we need to bring back a reasonable, workable version of the heart exam. The question is, will we do it?

Carol Pfeiffer is a tall, slender brunette with a husky voice and a warm smile. She is sitting at the head of a table in a small conference room crammed with a half dozen second-year medical students dressed in their short white coats. A few of the students sit; the others move restlessly around the room. They chat nervously as they wait. Tension fills the air like a bad smell. The students are there to take their end-of-the-year final but there are no blue books, no number 2 pencils, no desks. This exam consists of half a dozen simulated patient encounters.

The patients these students will be seeing are actually actors who have been trained to depict one or more of the 320 medical conditions on which the students will be tested. Carol is the head of the Medical Skills Assessment Programme at the University of Connecticut. She explains the test to the anxious students, even though these guys are old hands at this – they took a similar test at the end of their first year and have learned from these patient-instructors throughout their first two years.

The test is set up to simulate an outpatient doctor's practice. The students will visit the six rooms in the order given on each one's schedule. Outside the door there is a little card listing the patient's chief complaint. When the bell rings the students will enter the rooms and begin collecting the essential information on each patient. They will get the patient's history, perform a physical exam, explain to the patient what they think is going on. Once they leave the room they will write a brief medical note on the patient.

The rooms are equipped with the usual doctor's office stuff – a small table with a couple of chairs, an exam table, a blood pressure cuff, and thermometer – plus some equipment not usually found in an office – a small camera and a microphone. The entire encounter will be videotaped and the students and their teacher will review it after the test. After reminding the students about how the test works, Carol asks for questions. When there are none she sends them to the corridor around the corner, to find the room with their first patient.

I follow Pfeiffer into what looks like the control room of a TV studio. It's dominated by a wall of small black-and-white monitors. I don a set of headphones and plug in to watch one of the encounters. Most of the scenarios require the student to recognise a common illness and recommend the appropriate study or treatment. In one room there's a young man complaining of shortness of breath – his history reveals that he has had an accidental exposure at work to toxic chemicals. Diagnosis: asthma due to occupational exposure. In another room a fifty-something-year-old man complains of chest pain with any exertion for the past day. Diagnosis: likely unstable angina. Some need a diagnosis and counselling: a worried mom brings in her daughter, who has a cold and ear pain. She wants antibiotics for her little girl. The student's job is to explain why antibiotics are not appropriate. A young woman complaining of trouble sleeping is found to have a pattern of binge drinking, putting her at risk for alcohol-related disease and disability. The student's job in this case is to counsel the woman about the risks from her behaviour.

After checking in on a few of the rooms, I settle in to watch a young man who is speaking with a heavyset patient with greying hair. The student introduces himself and washes his hands as he's been taught. He sits and asks the man what brought him in. It's his stomach, the man tells Chris, the young doctor-to-be. Every now and then he gets this pain that comes on an hour or so after he eats. It doesn't happen all the time but a couple of nights before it woke him up from sleep and he almost went to the emergency room but decided to come in to get it checked out instead. The pain was severe and constant, lasting several hours. That time he thought he had a fever as well. Sometimes he has diarrhoea when he has the pain.

As the student asks questions, more details come out. He sometimes takes an antacid for the pain but it doesn't seem to do any good. The pain seems more common after a meal of fatty foods. The other night he'd had fried chicken. The pain seems to be mostly on his right side and doesn't worsen when he lies down; he's never noticed black or tarry stools, which would suggest a bleeding ulcer. The student gets the rest of the patient's history. He has high blood pressure and takes two medications for that; he's married, works in an office, doesn't drink or smoke. He's been on a health kick lately and lost twenty pounds over the past couple of months. The fried chicken was a little treat to celebrate his success.

Now it's time for the exam. The student, a beefy young man with light brown hair and an open pleasant face, asks the man to move to the exam table. The exam is perfectly normal until he gets to the abdomen. Chris presses gingerly on the right side, just below the rib cage. The man grunts in (mock) pain. He asks the patient to take a deep breath and as he's inhaling the student pushes briskly in the same area. The man grunts again. Chris tells the middle-aged man that he thinks maybe he has a gallstone and that the pain is caused when the stone blocks the duct leading out of the gall-bladder. He'll need to get some tests before he can confirm that diagnosis, he concludes somewhat vaguely. The student shakes the man's hand again and steps out of the room.

I watch on the monitor as the 'patient' opens a drawer and removes a form and a pen. He quickly moves through the yes/no answers by which he evaluates the student. Yes he introduced himself, and yes he washed his hands. No he didn't always use simple language. Yes he examined the abdomen. Yes he listened for the presence of bowel sounds and pressed on the right upper quadrant.

Suddenly there's another knock on the door and Chris walks back into the room. I forgot to do a rectal, he tells the surprised patient. Invasive exams such as this are not actually performed in these tests. Instead the student tells the patient he would like to do one and the patient gives him a card with the results of the exam written on it. But not this time. 'It's too late for you to ask for that,' the patient tells him. 'You're out of here.'

After Chris finishes up his note, he returns once more to the patient's room. The patient reviews how the student did in the encounter. He notes that Chris opened the encounter well but stumbled as he was asking questions about the pain. 'Don't worry about making sure you ask every single question on the list,' he tells the student. 'You know this material. Let your instincts tell you where to go with your questions.' And another point. 'Be sensitive to the patient. Once you have figured out where the pain is, don't keep pressing on the spot.'

After the test I sought out Chris as he was collecting his backpack from the conference room. The room was filled again but the difference was immediately apparent. The med students were laughing and talking about the mistakes they made. There was the giddiness of pressure relieved. 'The hardest thing is that you can't write anything down while you're in with the patient,' Chris tells me. 'You have to hold it all in your head. You know I kind of dread these exams but we all know we need it.' He's planning to go into surgery, but, he quickly adds, that doesn't mean he doesn't need to know how to do all this. 'Surgeons see patients at the office too.'

Certainly there is some pretty good evidence that these skills will come in handy no matter what area of patient care a doctor goes into. But these students will need to know the clinical exam well before they go into whatever specialty they have planned. At the end of their four years of medical school each of these students will be tested on these very same skills in the very same way.

Starting in 2004, all medical students in the US have been required to pass an exam that tests their clinical skills: their ability to take a history, perform an appropriate physical exam, and collect the data needed to diagnose and treat a patient. The United States Medical Licensing Examination – known as the USMLE – is the test physicians must pass to get licensed in most states. When I took the exam it was made up of just two parts. The first, given at the end of my second year of med school, tested knowledge of the basic sciences of medicine – anatomy, physiology, pharmacology, genetics. The second part of the test was given after graduation and focused on the understanding of basic patient care concepts – could I interpret the patient data that was provided? Was I able to formulate an appropriate differential diagnosis? What studies should be ordered based on what was known? Which medicines would be appropriate in the given setting? Which would be dangerous and must be avoided? Students must still prove their mastery of the book knowledge of medicine, but now, in addition, they will have to demonstrate their skill with patients as well.

In adding this component to competency testing, the USMLE is hearkening back to an older model. As early as 1916 the licensing exam included an evaluation of a real patient, observed by an experienced physician-grader. After taking a history and performing a physical exam, the students were questioned about what they found. This component was dropped in 1964 because of the lack of standardisation intrinsic to this kind of test.

But twenty years later the licensing board was asked to design a new test of these skills that would be reliable. The National Board of Medical Examiners, which oversees the USMLE, spent another twenty years trying to develop a system for testing these skills that was fair and reproducible. The medical school class of 2005 was the first to have to jump through this additional hoop.

Medical schools didn't exactly embrace this new test with open arms. The American Medical Association (AMA) was against it. So was their student branch as well as the student arm of the American Academy of Family Physicians. Opponents argued that most medical students already learn this stuff; and most institutions already test it, so what's the point of repeating this testing? To the students it seemed like just one more expensive test – they have to pay to travel to one of a dozen centres across the US, and the test itself cost over $1,000. But ultimately everyone takes it because that's what you need to do to become a doctor.

Has it done any good? It's still too early to tell if the test has made any real difference in what doctors do, yet if my own institution is any example, then I suspect it's having a tremendous impact on how doctors are trained – at least in medical school.

Eric Holmboe now heads the department that evaluates medical residents at the American Board of Internal Medicine (ABIM), the organisation that accredits doctors specialising in internal medicine. Until 2004 he was associate programme director of the Primary Care Internal Medicine Residency Programme at Yale. (That's when he saw my patient Susan Sukhoo.) At a recent meeting of directors of clinical teaching from medical schools in the Northeast, Holmboe described Yale's preparation for the clinical skills exam part of the USMLE. The faculty had arranged for all of the fourth-year medical students to go to the University of Connecticut in Farmington, where they could take the kind of test that Chris took as preparation for the real thing.

Before the test several of the Yale faculty travelled to northwest Connecticut to check out the facilities and the test. They chose seven clinical scenarios, giving them a few tweaks until everybody was comfortable with the setup. And students from Yale travelled up in groups of six to take the test over the course of several weeks.

When the scores came back, the faculty was shocked. Twenty per cent of these fourth-year Yale medical students – seventeen out of eighty-five test takers – had failed the test. Eric described the reaction when he presented the scores to the faculty. 'It was god-awful – the grief reaction in spades,' Eric told me. 'Kübler-Ross was hovering over the room,' referring to the anthropologist's famous stages of grief. 'It was anger, denial, and bargaining all rolled up in one.' There were concerns about the test – even though they had signed off on it before the students had gone up – and there was plenty of scepticism – this could not represent the real performance of fourth-year

Yale medical students. But amid grumbling and scepticism, everyone agreed to view the tapes of the students who failed.

When they met again, four weeks later, attitudes had changed. 'The anger and denial had evolved into deep, deep depression,' Eric reported. In one tape, a Yale medical student who was planning to go into neurology completely botched the cardiac exam. He was listening for heart sounds in all the wrong places. When he was given this feedback by the patient-instructor, the student's response was breathtaking in its arrogance and ignorance: he didn't need to know the heart exam – he was going into neurology. Stroke, the most common neurological disease, is often caused by problems originating in the heart. 'When he said that,' continued Eric, 'it pretty much cinched the deal and suddenly it was Houston, we've got a problem.'

In response, Yale revamped the way the physical exam was taught. When I was a student, the physical exam was taught at the end of the second year, just before we began our clinical clerkships that took us into the hospital wards. It was a twelve-week course with lectures a couple of times a week. During the lecture the physiology of the organ system was briefly reviewed and the exam technique was explained and sometimes (but not usually) demonstrated. Essentially I learned about the physical exam the way I learned about sex and menstruation – I got a brief, very nonspecific chat and a book. And did I have any questions? No. Great. The end. All the real info I was left to gather on my own. I figured it out at puberty and I figured it out again in medical school. Essentially I spent hours roaming the halls of the hospital looking for medical students already doing their clerkships to ask them to show me interesting physical exam findings. Like everyone I knew, I learned what I knew about the physical exam on my own, with a patient, a book, and the help and 'wisdom' of a student just one or two years ahead of me.

Now Yale begins teaching their medical students from day one. In the very first year there are classes on the techniques of interviewing and examination. Students meet in small groups weekly to review and practise these techniques for the first two years of school – first on each other, then on patients in offices and in the hospital. By the time medical students enter the hospital in their third year, they have the basics of these key data-collecting tools down. They are ready to build on a sound foundation. Unfortunately, there is frequently no one there to help them start construction.

I graduated from medical school with a set of physical exam skills that was spotty and idiosyncratic, and may have been considered unacceptable – had the doctors I then worked with ever observed me. I wasn't worried, though. I figured I'd learn the proper way to examine a patient when I was a resident. I was wrong. Studies show that by the end of residency training a physician's skill may be no better than the skills he had as a medical student.

Some of this is undoubtedly due to the time and access constraints already discussed. But some of this is due to an underlying attitude that the physical exam is already history. I accompanied Holmboe to a meeting with several directors from medical school and residency programmes to discuss a new initiative to shore up the clinical skills of doctors in training launched by the American Board of Internal Medicine (ABIM). At this meeting Dr Raquel Buranosky from the University of Pittsburgh voiced a common complaint. 'Med students in our programme get hours and hours of training in the physical exam in their first and second years. They do great at our final exam. Then they go into their clinical clerkships and, poof, it's gone.' There was general head nodding around the room and many of the directors told similar stories. Eric added one of his own. A colleague had worked with a medical student several times and been happy with his skills. Several weeks into the student's first clinical rotation – an internal medicine clerkship – the young student returned to have one last class with his teacher. The teacher watched him evaluate a patient and was horrified to see the student do absolutely everything wrong. He interrupted the patient's story, he asked closed-ended questions, he examined patients through their clothes. He skipped much of the exam. The teacher couldn't believe it. He asked the student what had happened since they last met. Oh, replied the student, 'my resident says we don't have time to do all that. I mean, what's the point?'

Anyone who's been through training won't doubt the accuracy of this young man's story. In residency, it often seems that no one cares if the patient is examined or not. Small wonder that many of the finer points of the exam simply slip away. And once they're gone, it practically takes a miracle to get them back. And yet with a patient like Patty Donnally, these skills can unravel a mystery.

## *A Kink in the System*

Patty Donnally is a youthful-appearing fifty-eight-year-old woman who has had high blood pressure since she was a teen. And no matter how many medications she's taken – and she has taken many – it's never been well controlled. Her internist tried for years to tame it. He put her on every combination of medications he could think of. Her blood pressure came down – but was never normal. Not even close. Occasionally he wondered if she was even taking her medicines. But she came to all her appointments, was aggressive in following up, even read up on her problem. That wasn't the behaviour of someone who didn't take her meds. And when asked, she could recite her most current medication regimen no matter how many times it had changed. No. It was clear – this lady took her medicines. But her blood pressure remained high. After almost a decade her internist gave up and referred her to a specialist in hypertension. The specialist was baffled too and eventually he referred her to the hypertension clinic at Yale.

At Yale she was seen by Dr Bill Asch, a young enthusiastic hypertension fellow whose cheerful disposition often made her forget the frustrations of her apparently untreatable disease. His wit made the schlep to New Haven almost worth it. So she was disappointed and a little annoyed when a new doctor walked through the door.

'Where's my regular doctor?' she asked the young woman who entered the tidy exam room. A trace of annoyance coloured her voice, and the lines between her brows deepened in the top half of a frown. Dr Shin Ru Lin sighed inwardly. She had finished her residency training a few weeks before, and had just started at Yale's hypertension subspecialty training programme. She was getting to know the patients she inherited from Asch, who was doing research this year and not seeing patients. Serious and shy, she'd been a little hurt by the disappointment expressed by more than a few of his patients when they found that she was now going to be their doctor.

And she was more than a little intimidated by this case in particular. Ms Donnally was on six potent hypertension drugs, and yet, according to the nursing note on the front of the chart, her blood pressure was still too high. The patient had seen many doctors, had had scores of tests. The chart was inches thick, and still no one understood what was going on. Lin had only just begun her graduate fellowship in hypertension – how was she supposed to figure this out? What could she possibly have to offer?

'When were you first diagnosed with hypertension?' the doctor asked tentatively.

'I've had it forever – you know, it's all in my records.' Patty waved toward the thick chart. 'My blood pressure is too high, I'm always tired, and my legs hurt when I walk. Nothing changes – except my doctors.'

In a specialty clinic like this one for hypertension at Yale–New Haven Hospital, patients have already been to several doctors, and often they are as frustrated as the physician who referred them. Each specialist, each series of tests, eliminates more of the likely causes of the problem, and the diagnostic question seems increasingly difficult to answer. And in an academic medical centre, patients are often seen by trainees, like Lin, who change every year.

Lin was overwhelmed. Waiting outside the exam room as the patient undressed for the physical exam, she opened the thick chart. She knew it would take her hours to go through it properly and she still had several more patients to see. Lin scolded herself for not reviewing it more thoroughly before meeting her for this first visit. She quickly paged through it. High blood pressure – okay. Also high cholesterol. She took a medicine for that. She didn't smoke or drink. She carefully kept track of her blood pressure at home. Before the doctor could get much further, it was time to go back in.

On exam, the patient's blood pressure was – as expected – very high. But there were unexpected findings as well. As Lin listened to the patient's neck over the carotid arteries, she heard a soft rhythmic whooshing noise over the normally silent vessels. This sound, known as a bruit, is caused by an unnatural turbulence in the flow of blood. It often indicates a narrowing of the arteries caused by atherosclerosis, commonly referred to as a hardening of the arteries.

She moved her stethoscope down to the chest. She heard more unexpected noises. In between the lub and dup of the normal heartbeat there was a brief, harsh murmur – like the snarl of an angry animal. Was this a new symptom? She would have to check the chart. It was audible everywhere she placed her stethoscope on the left side of the chest, though it seemed loudest at the top. Atherosclerosis could affect the valves of the heart as well as the arteries. This raspy murmur suggested that the disease may have narrowed the patient's aortic valve, one of the four valves of the heart. Could that be driving her blood pressure up? It seemed unlikely.

Then, in the abdomen, she found yet another noise: a soft shush-shush over the renal arteries. As she completed the exam, Lin remembered the patient's other complaint and examined her legs and feet. They looked fine – no lesions, redness, or rashes – but she couldn't find a pulse at either ankle. Was this more evidence of hardened arteries diminishing blood flow to her feet? That could explain the pain in her legs.

Finally, she asked herself the question all doctors must ask at the end of a visit: what could she do for this patient today? She added yet another medicine for the high blood pressure. And she would need to check the patient's cholesterol. Even though she was on one cholesterol medication, if all this noise and the leg pain were from narrowing of the arteries, it would be essential to bring her cholesterol down as low as possible.

What about the heart murmur? Although Lin couldn't imagine how a narrowed valve could drive the patient's blood pressure up, she thought it made sense to be thorough in a case this elusive. An echocardiogram would show whether the noise was coming from an abnormal cardiac valve.

That evening Lin sat down with the patient's chart. Before figuring out what she could do to solve this puzzle, she needed to know what had already been done. The most striking feature in this patient's case was a remarkably high level of renin, a chemical made by the kidney to increase blood pressure. When the kidneys receive too little blood, they release this enzyme, which increases blood flow to the kidneys by increasing the pressure in the arterial system – the way you might get water to a distant flower bed by increasing the pressure in a garden hose. This woman produced one hundred times the normal amount of renin. No wonder her blood pressure was abnormal.

So what in the world could cause the kidney to produce so much renin? Most commonly that occurs when atherosclerotic disease, the thickening and hardening of the vessels of the body, blocks the arteries supplying the kidney with blood. Perhaps that was the problem, she thought triumphantly. No, she realised moments later. An earlier angiogram had showed there was nothing blocking the arteries that carried the blood from the aorta to the kidneys.

Could she have a renin-producing tumour? There have been cases of these types of tumours in the kidney. No, an MRI of the kidney hadn't shown any tumours. Adrenaline makes your renin go up. Could she have an adrenaline-producing tumour? That had already been ruled out too. As Lin closed the chart and packed up to leave, she worried that she would have nothing new to offer the patient when she returned.

The following week, Lin ran into the attending doctor with whom she had seen the patient. 'Hey, Shin, did you see the results of the echo?' he asked, referring to the echocardiogram and brimming with excitement. 'Do you know what it showed?' He paused dramatically. 'Aortic coarctation.' Lin felt her eyes widen. She had found the cause of the hypertension – but that disease hadn't even crossed her mind. It was a diagnosis made by accident.

The aorta is the large, muscular vessel that takes blood from the heart and delivers it to all the parts of the body. A normal aorta is about three centimetres wide, about the size of a half dollar. In coarctation, the aorta develops abnormally, and instead of being a wide-open tube, it has a kink, narrowing the tube and limiting the flow of blood. The kidneys weren't getting enough blood, just as Lin and the other doctors had suspected. They had looked for such a blockage, but in the wrong places. Instead of being next to the kidneys, it turned out it was just inches from the heart.

Once Lin confirmed the diagnosis with an MRI, the patient was referred to Dr John Fahey, a cardiologist with experience in the delicate process of repairing the aorta. The day after her surgery, Ms Donnally told me, she needed only one medication to control her blood pressure. It was, she said, a miracle. And the pain in her legs diminished. Like her kidneys, the muscles in her legs must have been starved of blood.

## The Old/New Science of the Physical Exam

Why hadn't Lin, or any of the patient's previous doctors, considered coarctation of the aorta? If you look at a list of causes of hard-to-treat hypertension, it's always on that list. And yet it had been missed. Certainly it's an unusual cause of hypertension in an adult – mostly because it's ordinarily picked up in childhood. It's the number one cause of high blood pressure in children but far down the list of causes of adult hypertension. And yet doctors often think of diseases that are just as unusual. High on Dr Lin's differential diagnosis was a renin-producing tumour. An exceptionally rare disease. This patient had already been tested for this and other diseases even less common than coarctation.

Moreover, Donnally had all the classic signs and symptoms. She had the murmur that was heard throughout her chest, neck, and abdomen. She had no pulses at all in her lower extremities and pain in her legs when walking. And of course she had high blood pressure. Yet it was still missed, not by one doctor, but by many. I spoke at length with Dr Lin and Dr Asch about why this diagnosis was missed. Both confessed that they hadn't done the one physical exam test that would have most strongly suggested this diagnosis: comparing the blood pressure in the arms to the blood pressure in the legs. Normally the blood pressure in the legs is the same or higher than that in the arms. But because of the narrowing of the aorta, patients with coarctation provide less blood to the lower half of the body than normal. And because there's less blood, the blood pressure taken in the legs would be lower rather than higher.

When they finally checked, indeed the blood pressure in this patient's legs was much lower than that found in her arms. Both Asch and Lin say they now do that exam routinely on patients with resistant hypertension. But they didn't do it then. Of course since both doctors were in training, they were supervised in their care of this patient. Dr John Hayslett, a well-known researcher and hypertension expert, carefully reviewed the care given to each patient at Yale's hypertension clinic. His work has appeared in the most prestigious journals in medicine and his clinic at Yale is considered one of the best in the US. He never asked about this particular physical exam test. Says Asch, he probably assumed it had been done in the course of doing a thorough physical exam – if not by these fellows then by any of the dozen or so doctors who had already seen this patient.

Hayslett couldn't know whether this particular exam had been done because he hadn't seen these postgraduate fellows do the exam. The assumption is that by the time you get to this level of training, assessment of the basics – like the physical exam – is simply not necessary.

This is a common assumption, says Dr Eric Holmboe. 'We send a resident or medical student into a room with a patient, telling them to take a history and perform a physical exam. They come out and we ask them what they found. That's like sending a music student into a soundproof room with a piano and a piece of sheet music and asking them when they came out, So, how'd you do? It's crazy. How would they even know? You'd fire the music teacher who taught that way.' Maybe at some point in the past there was no need to evaluate the basic data gathering – though Holmboe is not sure there has ever been a time when teachers could assume these basic skills were done well. 'There's a tendency to think that back in some previous golden era things were better. I call that *Nostalgialitis imperfecta*,' he continued with a smile. 'But there's plenty of evidence that there were significant inadequacies in the way doctors took a history and performed a physical exam starting as early as the 1970s.'

Eric wants to change all that. An energetic man in his forties with a rangy build, broad smile, and loping gait, he greeted me enthusiastically when I appeared at one of his workshops taking place in Boston. Eric is in charge of developing programmes to shore up the physical exam training in medical residency programmes for the American Board of Internal Medicine. One of the principal ways he does that is by teaching teachers how to teach. His focus is to convince teachers to actually watch residents as they examine their patients and then teach them how to fix what they find. 'The way I was taught the physical exam was just crazy,' he told me. 'No one ever watched me. How could they help me get better? I could count on one hand the number of times I was observed performing the most basic parts of my job.'

When Eric finished his training in internal medicine at Yale, he returned to Bethesda Naval Hospital to complete his military service. His job was to teach residents who were training at the hospital. Fresh out of his own residency, Eric recalled his frustrations with the system and began observing residents on the job, as they evaluated the patients who'd come to the hospital or the clinic. At first the residents were anxious about his presence. No one had ever done this before. Some were worried that they were being singled out. Had Eric heard something that made him question their abilities? Over time, Eric was able to convince his residents that this was an important and useful practice for everyone in training – not just those with problems.

'It didn't take long before the trainees in our programme began to welcome these observed encounters. I wouldn't say they begged for them, but they were happy to have me there and I think they found the feedback extremely useful.' And, he continued, they needed it.

'I couldn't believe what these residents were doing. Examining people fully clothed. Listening to the heart and lungs through layers of clothing, placing the stethoscope in the wrong places. Poking, prodding, and thumping in places where it just won't tell them anything.' And he found residents almost universally grateful when he showed them a better way of doing it. 'The physical exam just becomes a much more useful tool when you use it correctly.'

In a paper first promulgating the use of direct observation as a tool in evaluating residents, Eric wrote: 'Direct observation of trainees is necessary to evaluate the process of data acquisition and care. A trainee's ability to take a complete history; perform an accurate, thorough physical examination;

communicate effectively; and demonstrate appropriate interpersonal and professional behaviour can best be measured through the direct sampling of these clinical skills.' It seems obvious and yet it's been a remarkably hard sell – not just to residents but to training programmes as well. It's time-consuming and many physicians are not comfortable enough with their own physical exam skills to feel competent to supervise the skills in someone else. And it simply wasn't the way things were done – traditionally.

That tradition is summed up in one phrase that I heard frequently in my own training: see one, do one, teach one. It's how residents have been taught to do procedures for decades. It also describes how many were taught the physical exam. A study published recently shows how inadequate this style of teaching is. A group of residents in nine teaching hospitals in England were asked to describe how they were taught to perform seven relatively simple procedures – from giving a shot to taking an EKG. They were also asked about their confidence in their own ability to perform this procedure the first time they did it. The same questionnaire was given to a group of nurses who traditionally get highly structured training in the performance of procedures. Over a third of doctors said they received no training at all before performing the procedure and nearly half said they felt unqualified when they first performed them. Nearly half performed these procedures unsupervised when they did it the first time. Doctors are often sent out into the wards to perform on their patients – with inadequate training, and sometimes no training at all – procedures that carry some, usually small, risk to the patient if done incorrectly. And yet we continue to allow medical students and residents to perform these procedures without adequate training. The same is true of the noninvasive clinical stuff – the taking of a history or the performing of a physical exam that doctors do where there is no risk of directly doing harm, only of missing something important.

So Eric has spent the past several years as a one-man sales force, travelling from training programme to training programme selling the idea that direct observation of residents in training is the right thing to do. He has developed a four-day course to teach the teachers how to observe. One of the problems, says Holmboe, is that since many doctors themselves were not given any formal training in these skills, most doctors haven't developed formal criteria of how to talk to a patient and how to examine a patient. If doctors aren't certain that they are doing it right, how can they know if a student is doing it right? As a teaching tool, Eric scripted and videotaped three clinical encounters where a resident was shown taking a history, performing the physical exam, and counselling the patient. He taped three versions of each of three scenarios: one of poor quality, one of moderate quality, one of high quality. Then he asked teachers to grade each encounter. The grades were all over the map. The encounters that were poor were given grades as high as the high-quality encounters. No one had a clue. This class helps teachers develop criteria for each component of the clinical exam and teaches them to apply them when they watch a trainee. Teachers are also coached in how to provide feedback in a constructive and useful way.

There are over eight thousand residency programmes in the United States, and Eric hopes to reach them all. How well does the programme work? Certainly physicians who complete Eric's programme say that they feel much more comfortable watching residents and giving feedback. Whether better teaching translates into better doctoring is still unknown. But Holmboe is travelling to as many as he can in a one-man effort to resuscitate the physical exam. And yet Eric remains hopeful. His optimism engenders a little of my own. Maybe he can do it after all.

# PART THREE

## *High Tech*

## CHAPTER EIGHT

## *Testing Troubles*

Carol Ann DeVries felt like she was falling apart. A compact woman with a cheerful, round face and deep-set brown eyes, she had been healthy all her life. Then, just a few weeks after her fifty-ninth birthday, everything changed. Out of nowhere she got a rampant case of hives. A short course of prednisone cleared them up, but neither Carol Ann nor her internist could figure out where they'd come from.

Then, one Saturday morning, a few days later, she awoke feeling achy and hot, her throat was sandpaper, and she had an odd red rash near the base of her spine. Was this more hives? Carol Ann had a doctor's appointment scheduled for the next week, but she felt too awful to wait. She drove herself to the emergency room of her local hospital.

The ER doctor took her temperature, looked at the rash, and briskly told her she had Lyme disease. 'An antibiotic will clear it up,' he said, scribbling the prescription. 'One pill twice a day for two weeks,' he told her, and he headed out the door. 'Wait a second,' Carol Ann called after him. 'Aren't you even going to get a test to see if I have Lyme?'

'You don't need it,' he told her, ticking off the items that supported his diagnosis. It was early summer, when Lyme is most common. She lived in suburban Connecticut – not too far from the actual town of Lyme, where the disease was first identified. And she had a big, round rash typical of those seen in the early stages of Lyme disease.

He acknowledged that her symptoms weren't the classic headache and stiff neck, but, still, she had the fever and body aches. The odds were overwhelming that this was Lyme, he told her. 'Besides, this early in the disease, the Lyme test wouldn't tell us a thing.' Then he was gone, off to the next room, the next patient, leaving Carol Ann with his scrawled prescription and a feeling of uncertainty.

Every spring and summer some version of this story is repeated tens of thousands of times in states of the Northeast, Midwest, and northern West Coast. Often, as in Carol Ann's case, the diagnosis will be made without a test, based on the patient's geography and symptoms, and cinched by the presence of the typical rash, known as erythema migrans. The diagnosis will be appropriate and reasonable, but not definitive. And in Lyme disease, that uncertainty has proved to be a particularly noxious ingredient.

Carol Ann took the antibiotics as prescribed. By the following weekend she felt almost back to her usual self. For most patients with Lyme disease, a single course of antibiotics is curative. But if Carol Ann had had a simple case of Lyme disease, I wouldn't be telling you this story. Instead, a few weeks later, Carol Ann developed pain and stiffness in her knees and hips. There was no swelling, no redness, just this strange reluctance in the joints of her lower body.

She went to her internist, who thought the symptoms were from Lyme disease. Untreated or inadequately treated, Lyme can attack the joints, causing pain, and usually swelling. He changed her to another antibiotic – doxycycline. She stayed on that for three more weeks but the stiffness continued. Her internist was baffled; he sent her to a rheumatologist. The rheumatologist wasn't sure what was going on either. So she went back to her internist. 'He fobbed me off on his physician's assistant,' Carol Ann said. 'I was practically crying over the phone because of the pain. I told the PA I couldn't even sleep because the pain was so bad. She wasn't very sympathetic.'

Carol Ann felt abandoned. Her doctor was a nice guy, she told me, but he clearly didn't know what was causing her pain or what to do about it. She decided to take matters into her own hands. She talked to friends; she rummaged the shelves of her local bookstore; she cruised the Internet.

Everything seemed to point her back to her diagnosis of Lyme disease. She decided that she needed a Lyme specialist – someone who really understood the disease. So she set off to find one.

What Carol Ann didn't know – couldn't know – was that she was about to enter one of hottest controversies in medicine, a maelstrom of professional contention and confusion about Lyme disease, a controversy that would leave her in pain for two full years. Most patients believe that it is usually possible to determine what ails them with some kind of test – an X-ray, for example, or an MRI, or any of the hundreds and hundreds of blood and urine tests. If the test is positive, the patient may not be happy, but at least he believes he knows exactly what's wrong: a fractured wristbone, asthma, a tumour, a heart attack. If the test is negative, then the patient believes the result at least proves he *doesn't* have something, which can be a relief if what he thought he had was cancer or some other terrifying disease. Or it can be very frustrating – since often treatment and the possibility of cure depend on diagnosis.

Doctors too put a great deal of faith in the power of diagnostic tests. And mostly for good reasons. Tremendous strides have been made in our ability to identify a disease using advanced technologies of one kind or another. While the patient's story and physical exam can often suggest a diagnosis, both doctors and patients like to see hard evidence – and that evidence usually comes in the form of results from some kind of diagnostic test.

But, as it turns out, tests and their results are not nearly as crisp and clear as many patients (and doctors) assume them to be. In fact, for all of their tremendous and invaluable power, the testing process can actually slow or sidetrack the diagnostic process in some cases – or derail it completely.

The complexities surrounding testing for Lyme disease have pitted doctors against doctors and led to a confusion bordering on chaos about the diagnosis of this common and highly treatable disease. The result has been a virtual epidemic of missed and mistaken diagnoses. Some patients end up suffering from undiagnosed acute Lyme disease. And hundreds – maybe thousands – of patients sick with other diseases are being 'diagnosed' with a phantom illness and treated for a medical problem they don't have.

To Carol Ann it made sense that these aches and pains could be linked to her Lyme disease – after all, she'd been fine until then. Plus, her doctor had thought the first round of antibiotics ineffective. Why would the second round be any better? She finally located a Lyme specialist in nearby Wilton, Connecticut. He didn't take insurance – none of the Lyme specialists she called did – but his fee was reasonable and he was conveniently close.

By the time Carol Ann went to see the specialist, Dr Matthew Davidson (not his real name), she was a wreck. Her body ached all the time. The joints didn't look injured – there was no swelling or redness – but they hurt so much that even sleeping was difficult. She was exhausted, her memory was shot, she couldn't concentrate, and normal daily frustrations often reduced her to tears.

Davidson is a general internist who has focused his practice on Lyme disease. A stocky man with thinning blond hair, he exuded a warmth and openness that impressed Carol Ann. She sat in his exam room and began to describe the symptoms that had taken over her life for the past year.

Davidson nodded his head as Carol Ann began to list her symptoms. Her illness was no mystery to him. Her presentation was classic, he told her halfway through her story. What happened to her was common – far too common, in his opinion. She was right to seek his help. The antibiotics obviously hadn't worked; she hadn't been cured of her infection, and as a result she now had something he called 'chronic Lyme disease.'

Davidson explained that often an initial course of antibiotics doesn't kill the bacteria that cause Lyme disease. The bacteria somehow manage to 'hide' in the body, only to reappear and cause a host of symptoms that include joint pain, muscle pain, insomnia, and lack of concentration – all symptoms that Carol Ann had. Davidson said that her only hope was to take even more antibiotics. Maybe for a few more weeks, possibly for months, perhaps even for years, in the hope of finally eradicating the insidious bug and its symptoms. It could be a long process, he told her, but with his help she could defeat the infection and regain her health.

As Carol Ann left Davidson's office that day, she felt more optimistic than she had since her symptoms had started. That mood wouldn't last long.

The diagnosis that Carol Ann was given – chronic Lyme disease – is one that tens of thousands of patients have been given in the thirty-some years since Lyme disease was first identified. And a whole cadre of doctors like Davidson, who call themselves 'Lyme literate,' assert a special expertise in what they claim is a chronic and insidious infection. But despite the claims of these physicians – and the sometimes fervent belief of their patients – 'chronic Lyme disease' is almost certainly a phantom illness. Contrary to the claims of doctors like Davidson, there is little evidence that the bacteria that cause Lyme disease can persist in the face of antibiotics, causing the symptoms attributed to 'chronic Lyme disease.' Furthermore, there is plenty of solid evidence that shows that long-term treatment with antibiotics will *not* cure whatever it is that ails those diagnosed with this syndrome.

Despite this evidence, thousands of patients continue to be prescribed months and sometimes years of multiple antibiotics in a desperate quest for relief. The danger of this diagnosis and treatment are twofold. First, it puts patients at risk of serious side effects from the powerful drugs that are used. Second, this erroneous diagnosis can postpone diagnosis and treatment of other diseases, leaving patients even worse off than when they started.

How can reasonable, well-meaning medical doctors such as Davidson continue to believe in this phantom and continue to prescribe treatments that don't work? The answer has to do, at least in part, with the difficulty of diagnosing this complex disease. But it also is closely linked to a very human discomfort with the uncertainty when faced with a patient in pain and in need of an answer.

## *The Discovery of Lyme Disease*

The discovery of Lyme disease is one of the great pieces of medical detective work of the twentieth century. In 1956, Polly Murray, a young artist and housewife in Essex, Connecticut, began to suffer an array of inexplicable health problems: fevers, rashes, joint pains, and fatigue. Her memory didn't seem as sharp as it once was. She felt unfocused and found it hard to concentrate on her artwork. She went to her doctor. He was baffled. So were the specialists she was sent to see. Several suggested that the symptoms were all in her head – a manifestation of some psychiatric illness.

By 1964, Polly, her husband, and four children had moved to the small town of Lyme, Connecticut, an affluent community wedged into the verdant countryside between the Connecticut River and Long Island Sound. By then everyone in the family was suffering from the same symptoms Polly had. Even the dog was afflicted. Visits to the doctor were frequent, relief was rare, frustration high.

Over time Polly discovered that other people in the area were experiencing the same constellation of symptoms. Together, sufferers in her town had racked up hundreds of doctor visits and seen dozens of specialists. No one had an answer. No one could explain what they had or why so many of them had it. Finally, in October 1975, Murray called the state health department to report the strange local epidemic.

The health department turned to Dr Allen Steere, a Yale rheumatologist who had spent his first two years out of medical school working for the Epidemic Intelligence Service (EIS), the investigative arm of the nation's primary public health watchdog, the Centers for Disease Control (CDC) in Atlanta. Steere asked Murray to come to his office in New Haven and bring her notes. Unlike many of the doctors Polly had seen, Steere showed a profound interest in her story. He collected the names of other people she knew who were suffering. Steere called each family on her list. They, in turn, gave him additional names, and eventually he compiled a list of twelve adults and thirty-nine children who had the same collection of symptoms as Polly and her family.

Steere immediately noted that the individual cases resembled juvenile rheumatoid arthritis. But that was a relatively rare disorder. What, he wondered, could have caused the clustering of so many cases of this uncommon disease in such a small area? He set out to discover what, if anything, the fifty-one individuals had in common.

The outbreaks were seasonal, reaching a peak each summer and then again in the autumn. Based on that, Steere quickly focused on the possibility that this was some kind of insect-borne disease. But few patients remembered being bitten. And those who did described the appearance of the bite differently. It took two years of hard work before Dr Steere and his colleagues fingered a culprit. Steere remembers the day clearly: it was in the summer of 1977 when a young man walked into his office with a vial containing a tick he had found after a walk through the woods near his home. The hiker had never seen one like it. Neither had Steere. It turned out to be an immature *Ixodes scapularis* tick, a tiny black-legged arachnid, new to the region. Local insect census takers had been tracking the tick's march across Connecticut. Comparing the location of Steere's mystery cases and the areas invaded by the tick produced a geographical match.

The final piece of the puzzle remained a mystery until 1981, when Willy Burgdorfer, an entomologist for the National Institutes of Health, identified the corkscrew-shaped bacteria transmitted by the ticks that actually cause Lyme disease. It was a new bacterial species – and it was named in his honour: *Borrelia burgdorferi.*

The *burgdorferi* bacteria normally live in the blood of deer and various rodent species. As a larva, the *Ixodes* tick (commonly called a 'deer tick') takes a meal of blood from its animal host and, if the animal harbours the bacteria, gets a dose of *burgdorferi* at the same time. The bacteria don't seem to bother the ticks. They just live quietly in the tick's gut.

The tiny arachnid has three life stages – larva, nymph, and adult. At each life stage the tick takes only one blood meal. Most cases of Lyme are transmitted in the tick's second stage. At that point the *Ixodes* tick is dark and tiny – about the size of a pinhead – and the tick can easily latch on to a passing human and find a quiet spot to attach itself and take its blood meal. In order to transmit the infection the tick needs to stay attached for at least twenty-four hours – which it can often do easily because it is so hard to see and the bite is usually painless.

The earliest and most common symptom of Lyme disease is an expanding circular rash, which usually appears around the site of the tick bite within a few weeks. The rash often presents in a 'bull's-eye' pattern: a red ring circling a central clearing. However, some studies suggest that the most common presentation is a completely red, somewhat round patch that expands over the course of several days.

Left untreated, the *burgdorferi* bacteria migrate to other parts of the body and the immune system responds with inflammation, fever, muscle aches, and other symptoms as it tries to fight the infection.

Until Willy Burgdorfer's discovery of the bacterial cause of Lyme disease, no test existed for it – for the simple reason that nobody knew what to look for. Even after the identification of the bacteria, testing for the disease remained difficult. Many types of bacterial infections can be diagnosed by culture – taking a sample (a throat swab, for example), rubbing it on a material that fosters bacterial growth, incubating the sample for a period of days, and then identifying the colonies of bacteria that form. But the Lyme bacteria don't grow well in culture.

Instead, doctors seeking to diagnose Lyme disease must depend on the body's response to the infection. To do this, doctors use two separate tests, neither of which is good enough to use alone but which, used together, can reliably identify those who have been exposed to the bacteria. It's an old strategy and while once commonplace, it has been replaced by better, more specific tests in many diseases. It's still used for a couple of diseases such as HIV and hepatitis C – other diseases that cannot be easily grown in culture.

The first of the two tests is known as an ELISA (enzyme-linked immunosorbent assay) and it detects antibodies to the invading bacteria or virus in a person's blood. Antibodies are part of the body's defense system and help kill invaders. This ELISA looks for antibodies to the Lyme bacteria. And it's a pretty good test but cannot distinguish between *Borrelia burgdorferi* and many of its look-alike brothers, nephews, or distant cousins. Some types of normal flora can cause a positive reaction on the ELISA.

That's why a second test is needed. If a person tests positive or equivocal on the ELISA, then a second test is conducted called the Western blot test. Again, this test is looking at antibodies, not actual bacteria. This test looks for antibodies not to the whole organism but to the basic building blocks of the Lyme bacteria – individual proteins. It's a complicated process because many types of bacteria use the same building blocks. So it's not enough to identify, for example, two or three of the proteins known to be part of *burgdorferi* bacteria. Those same proteins are also found in many other species.

The CDC has determined a standard for interpreting results from the

Western blot test series. The standards say that Lyme disease should be suspected only if a person's blood is found to have antibodies to five out of the ten proteins that are commonly tested for. If a patient tests positive on ELISA *and* is positive on at least five out of the ten relevant Western blot tests, then it is very likely he has been exposed to Lyme disease.

If this were the end of the story, it wouldn't be so bad. The tests for Lyme disease would be indirect and require two steps, but the end result would be about as clear as we get in medicine. Unfortunately, it's more complicated than that.

First, it usually takes the body several weeks to develop enough antibodies against the bacteria to be measured by either of the two tests. In the earliest days of the infection, therefore, even though you may have the rash or other symptoms of early Lyme disease, neither of the two tests is likely to be positive. And if treatment is started early enough, the bacteria will be killed so quickly that the antibodies may *never* get created. That means there's no way to definitely prove, years later, that a person did *not* have Lyme disease at some point in the past. But an even more important complication in testing for Lyme disease is that once the body does make antibodies, they stick around – for months or years – as protection against future infections. That means that the tests for Lyme will remain positive long after the bacteria that originally caused the symptoms are eradicated. It will look for all the world as though you still harbour the bacteria even though you don't, because the tests don't measure bacteria – they measure antibodies.

### *The Phantom Epidemic*

Carol Ann, of course, did not understand these layers of complexity about testing for Lyme disease when she consulted with Dr Davidson. All she knew was that when she had left Davidson's office on her initial visit, he said he would send off a sample of her blood for a confirmatory test – which certainly seemed to Carol Ann like responsible medical practice. Two weeks later, the results came back: they were negative. This struck Carol Ann as odd, but didn't seem to bother Dr Davidson at all. He told her that testing was notoriously unreliable in making this diagnosis. He sent off more blood and started her on an antibiotic despite the negative test.

Several weeks later Davidson told Carol Ann the results of the second test were positive. He did not tell her that he wasn't following the guidelines for interpreting the results set by the CDC. Rather than the standard of five out of ten antibodies that the CDC regards as evidence of Lyme disease, her test had been positive in only three – which Davidson interpreted as 'positive.'

Like his 'Lyme literate' colleagues, Davidson justifies his more lenient testing standards as necessary for not missing anyone who may have the disease. But this is a poor argument. It's like saying that all patients who have a sore throat, runny nose, and a fever have the dreaded avian flu. You probably won't miss any cases of avian flu with such a vague set of criteria, but most of the time your diagnosis will be wrong. Instead most of the patients you identify this way will have other, far more common illnesses – a cold, maybe bronchitis, or maybe the usual flu.

But all this was invisible to Carol Ann, who took some comfort in the apparent confirmation of her illness. In any case, she was feeling a little better since starting the latest round of antibiotics. Her shoulder didn't hurt quite as much and she was sleeping better. But the medication was taking a toll on her stomach. She lost a few pounds because she felt nauseated – especially in the hours right after taking the medication. Davidson thought it was important to continue the antibiotic, especially in light of the 'positive' Lyme test, and encouraged her to soldier on. The hope of a complete recovery and of being her old self again made Carol Ann determined to continue taking the medication despite how ill it made her feel.

At about the same time that Carol Ann was forging ahead with her prescribed regimen of antibiotics, forty-four-year-old Will Hammer was negotiating slippery November roads to see his own 'Lyme literate' physician. He had been diagnosed with Lyme disease over a decade earlier but, when I spoke with him, he said he'd suffered from the disease for more than half his life. A tall man with short-cropped red hair and a quiet manner, he told me proudly that he hadn't missed a day of work because of his 'chronic Lyme disease' in over five years. He attributed his success to Dr Andrea Gaito, a rheumatologist and leader in the 'Lyme literate' movement who had had him on daily antibiotics for nearly thirteen years.

Hammer said he first developed symptoms in high school following a camping trip. He never had the fever, muscle pain, and headache typical of Lyme, but felt tired and run-down. 'Occasionally I'd feel not quite right and I'd wonder about it,' he told me, 'but it wasn't overwhelming.' The symptoms worsened in his twenties. He would have periods of insomnia, body pains, and fatigue. He went to doctor after doctor. No one even had a name for the illness he suffered, much less a cure.

Eventually he heard about Lyme disease and wondered if that could be causing his symptoms. He'd been tested for it in the past and been told the test was negative, but now he was hearing that the test wasn't very reliable. That's when he ended up in Gaito's office, who quickly diagnosed him with 'chronic Lyme disease' and started him on a six-week course of antibiotics.

The effect was immediate and life-altering, Hammer said. 'The first time I was treated here I felt better than I had in my entire adult life. I felt great.' But, he told me, that good feeling didn't last long.

A couple of months after completing his treatment, he started having trouble sleeping again. Then the fatigue and body aches reappeared. Then he began to have problems with his memory. 'At first it was little things. Then one day, I was driving my son to soccer as I had every weekend for months and suddenly I couldn't remember where to go. I couldn't quite remember where I was.' He pulled off the road into a parking lot. His heart was racing. Slowly and carefully, he figured out how to get where he needed to go. His young son, confused by his uncharacteristic behaviour, asked what was the matter. Nothing was wrong, he reassured his son, but inwardly he wondered what the hell was happening to him.

He went back to Gaito and started back on antibiotics. He's been on them with only a couple of short breaks ever since. At several points over the years since starting these medications, Hammer found himself as sick as he'd been to start with despite the daily antibiotics. He'd go back to his doctor – discouraged, frustrated, and depressed. Gaito would help him get through it, increasing the dose of the antibiotics or changing him to a new one and eventually he'd start to feel a bit better. When I last spoke to Hammer, he was going to Dr Gaito every three to four months, but he worried that he might never be well enough to stop taking the antibiotics.

Both Carol Ann and Will tell versions of a common story in Lyme country: a diagnosis of Lyme disease, followed by antibiotic treatment, an initial improvement, and then a return of symptoms. This pattern emerged early in the history of the disease. Allen Steere noticed that while most of his patients got well after a two-to-four-week course of antibiotics, between 10 and 20 per cent took months and occasionally even years to feel completely better. Like Will Hammer and Carol Ann DeVries, these patients often noticed a lingering fatigue and body aches. Some complained that they had trouble sleeping or problems with their memory. Still others would have recurrences of the joint pain and swelling that brought them to the rheumatologist in the first place. Steere dubbed this phenomenon Post–Lyme Disease syndrome.

In the early 1990s, Steere and researcher Nancy Shadick set out to determine how common the syndrome really was. They recruited one hundred residents from Ipswich, Massachusetts, an area that had been hard hit by Lyme disease. Half the patients had a documented history of Lyme disease, which had been treated; the other half had never had the disease. Nearly one in three of the treated patients continued to have residual pain and other symptoms more than five years after getting Lyme disease. These symptoms were seen far more frequently in those with Lyme than in those who'd never had it. Other studies too have found that those with a history of Lyme disease report more joint pain, fatigue, and memory problems than those who have never had the disease.

Initially there was concern that these symptoms represented an ongoing infection, which persisted despite a full course of antibiotics. Patients themselves said that it *felt* like an infection, and so Steere, Shadick, and many other doctors responded initially by treating these symptoms with a second or third course of antibiotics.

But it soon became obvious to Steere and others that while many of those with these persistent symptoms got better after multiple courses of antibiotics, so did people who only got the one round of antibiotics – those prescribed at the time of their original diagnosis. Ultimately, most people got better and it wasn't at all clear whether the repeated courses of antibiotics made any difference at all, once the disease had been treated initially.

To better understand what was going on, researchers in the field did what researchers do – they set up experiments to study in a very careful, controlled way whether antibiotics really helped people with Post–Lyme Disease syndrome recover faster.

The *New England Journal of Medicine* published the first experimental results in 2001. Researchers at Tufts Medical Center in Boston and Yale–New Haven Hospital looked at 129 patients who had documented proof of a previous infection with Lyme disease and persistent symptoms even after being given the recommended regimen of antibiotics. Most had some degree of musculoskeletal pain. Half were treated with ninety days of antibiotics and half were treated with an identical-looking placebo. Neither doctor nor patient knew who was getting which. Each participant was evaluated for physical and psychological health before, during, and after treatment with the drug or placebo.

Just over 40 per cent of the patients treated with the antibiotics felt better after the first month. And almost that many felt better, overall, after the full course of antibiotics and three months later. What about those on the placebo? The response was almost identical: 35 per cent of the patients getting a completely inert substance felt better after the first month and 40 per cent felt better by the end of the study. The antibiotics had made no difference at all.

Two other rigorous trials investigated the same issue. One showed a small improvement in symptoms of fatigue in those getting the antibiotics, but nothing else. The third study was done by psychologist Brian Fallon, a 'Lyme literate' researcher and the head of a research centre at Columbia University that is funded in part by a 'chronic Lyme disease' advocacy group. But even Fallon failed to find any significant difference between the group on antibiotics and the group getting the placebo. Moreover, in each of these studies a significant number of participants had complications stemming from antibiotic treatment. Many experienced the kinds of adverse reactions felt by Carol Ann, and sometimes the complications were serious enough that the study participants had to be hospitalised.

These kinds of consistent results from three separate, rigorous experiments, published in high-quality, peer-reviewed journals, would normally be considered a slam dunk in the medical world. The results clearly showed that antibiotics were *not* helpful for patients with persistent symptoms after being treated for Lyme disease, which strongly suggested that whatever was wrong with these people, it was *not* because they were harbouring some 'super *burgdorferi*' in their bodies. Leading medical groups involved in the study of Lyme disease published guidelines recommending against the use of multiple courses of antibiotics to treat the persistent symptoms. Normally, this would settle the matter and medical science would move on to a new mystery.

But that's not what happened here. Instead a small but vocal group of doctors and patients refused to accept these results, refused even to accept the designation of Post–Lyme Disease syndrome. They clung, instead, to 'chronic Lyme disease' and insisted that these symptoms *did* reflect an ongoing infection that warranted continuing treatment with antibiotics. They countered the randomised controlled trials with research of their own, which often showed improvement in patients given antibiotics. But none of these studies compared the antibiotics against a placebo. The randomised controlled trials showed that while patients getting antibiotics did improve, so did those getting the saltwater placebo. Studies done without the placebo had no way of telling whether the antibiotics were really effective or if the improvement was due to something in the normal ebbs and flows of any human condition.

Advocates of the 'chronic Lyme disease' model of understanding also seized on the complexities of Lyme tests and argued that because of their limitations, they should simply be disregarded. According to the International Lyme and Associated Diseases Society (ILADS), a group founded in 1999 to promote the dissemination of information on 'chronic Lyme disease,' diagnosis should be based *only* on the patient's symptoms. They don't trust either physical exams or the two tests for Lyme disease.

'Available data suggest that objective evidence alone is inadequate to make treatment decisions,' say the ILADS guidelines, 'because a significant number of Chronic Lyme Disease cases may occur in symptomatic patients without objective features on examination or confirmatory laboratory testing.'

The same guidelines assert that when the two-step testing for Lyme is used as recommended by the CDC, up to 90 per cent of cases are missed. But this is misleading at best. The 'research' offered in the guidelines to support this 'fact' consists of a single unpublished study. And if you use the test on everyone with the common symptoms of fatigue, insomnia, and muscle pains, 90 per cent of them probably will have a negative test because they don't really have Lyme disease.

In fact, when used as recommended, the tests accurately diagnose Lyme correctly more than 90 per cent of the time.

Instead of testing or the physical examination, ILADS and 'Lyme literate' doctors suggest that diagnosis be made on symptoms alone. The problem here is that the symptoms these doctors have chosen to rely on are hopelessly broad and over-inclusive. They include fatigue, sore throat, muscle pain, joint pain, insomnia, chest pain, abdominal pain, dizziness, nausea, poor concentration, headaches, irritability, depression, back pain. These are some of the most common symptoms of patients presenting to a primary care office. As an internist who sees patients regularly, most of the patients I see complain of these symptoms – and they certainly don't all have Lyme disease. Because I have an urban practice, I see only a few cases of Lyme disease a year.

And yet this is the diagnostic strategy that allows doctors like Davidson and Gaito to maintain that patients like Carol Ann or Will have 'chronic Lyme disease' and that they should be continually treated with antibiotics because of a lingering infection with the Lyme bacteria.

But there is probably an additional factor at work in the stubborn refusal of the 'Lyme literate' doctors and patients to accept evidence that most other doctors find compelling. It's an unavoidable aspect of all medical practice and is particularly related to the limitations of diagnostic testing – limitations that are particularly acute with Lyme disease. I'm talking about a discomfort with uncertainty, with ambiguity, with not knowing. And the doctors most uncomfortable in this way are the ones most likely to seize upon a diagnosis or diagnostic label and distort their own thinking in an attempt to prove to themselves and their patients that they know what's going on.

The fact is that, more often than doctors like to admit, they cannot find a cause for a patient's symptoms. This was powerfully illustrated in a 1998 study of two hundred patients diagnosed with Lyme disease. It turned out that more than half of the patients had *no* evidence of ever being infected with Lyme disease. As we've seen, it could be that some of these patients were treated for Lyme so quickly that they never developed the antibodies that would prove their infection. Maybe. But that surely could not account for such a huge fraction of these patients who tested negative for Lyme.

Only 44 of the 200 patients (20 per cent) were found to have active Lyme disease – with symptoms, physical exam findings, and blood tests consistent with Lyme disease. Another 40 (19 per cent) were found to have Post–Lyme Disease syndrome – with a clear history of Lyme infection, adequate treatment, and persistent symptoms. The other 116 patients in the study – 60 per cent of those enrolled – had no evidence of either current or past Lyme disease, even though all had been diagnosed with the disease. Clearly these results show that Lyme disease is being seriously over-diagnosed. But the results also shed light on the issue of uncertainty in medicine.

If the 116 patients who thought they had Lyme *didn't,* then what *did* they have? A lot of things, it turns out. Fourteen had rheumatoid arthritis. Fifteen had osteoarthritis. Eight were diagnosed with another infection, and another dozen had some kind of neurological disorder such as multiple sclerosis or amyotrophic lateral sclerosis (ALS), often referred to as Lou Gehrig's disease. Several were diagnosed with depression.

These diagnoses covered about half of the 116 – and clearly demonstrate how the phantom diagnosis of 'chronic Lyme disease' is concealing a great many real diseases that patients should be getting help with. But the other half of the participants are also interesting. These people were obviously suffering from real ailments – real symptoms – but neither the study doctors nor their own physicians could determine a cause. These patients had what doctors call 'medically unexplained symptoms.'

Nobody likes not knowing, but doctors, perhaps, find this state of being even more intolerable because it utterly thwarts their ability to alleviate suffering, which is often the fundamental motivation for their entire career. But a doctor's discomfort in the face of the unexplained can lead them astray. Rather than accepting a patient's symptoms as real, but unexplained, physicians all too often either dismiss the symptoms as unreal ('all in your head') or, alternatively, read too much into scanty evidence in an effort to banish uncertainty with a crisp diagnosis. Neither response serves patients.

We can see the results of both types of responses in Will Hammer's experience.

'My doctors would let me list my symptoms and they'd send me off for a bunch of tests and when they all came back normal they were like "What are you still doing here? We can't find anything wrong with you so it must be all in your head." But these pains I have, this fatigue, this confusion, it's not in my head. It's in my body. I needed someone to help me with what was happening to me. That's when I found Dr Gaito.'

By dismissing his symptoms, Will's doctors, in a very real sense, abandoned him. He didn't have a disease they knew about and so he wasn't really sick. This left him vulnerable to a wide range of other practitioners – both traditional and 'alternative' – who offer a sympathetic ear, a ready explanation for symptoms, and a confident plan of treatment. These are precisely what Dr Gaito provided. She listened to Will, validated his feelings, and offered him a compelling narrative of his symptoms – a seemingly solid and substantial diagnosis.

'Dr Gaito believes that this is chronic Lyme disease,' he told me. 'She's been treating me for it, and while I'm not all better, I shudder to think what I'd be like if I hadn't been taking antibiotics all these years. No, I think I'll probably end up taking antibiotics for the rest of my life and if that's what it takes, I'm willing.'

## *Escape from the Phantom*

Will's story is the result of a tangled web of factors, at the heart of which lie the unusually complicated and uncertain diagnostic tests used – or *not* used – for patients with Lyme disease. But whereas Dr Gaito and Will remain solidly wedded to their belief in the phantom 'chronic Lyme disease,' Carol Ann eventually escaped this psychological and medical dead end.

For weeks Carol Ann took the medicine prescribed by Dr Davidson. Nausea continued to plague her, but she persisted. Then her symptoms, which at first seemed to be getting better, slowly returned. Davidson changed her to a different dose of the antibiotic and when that didn't help, to yet another one. With each change, Carol Ann would start to feel a little better, but it never lasted. After months of this Carol Ann returned to Davidson's office – frustrated and sick. All of her symptoms had come back and the new medicine was making her feel just as nauseated as the old one had. At that point she'd been on antibiotics for over six months and she was – if anything – worse off than she had been when she first came to see him.

'I'm only sixty and I feel like an old woman,' she told him. 'What is wrong with me? If this is chronic Lyme disease, why am I getting worse?' It's an interesting question, he told her. If this were a persistent infection, he would expect her to get better. So – and he paused – maybe this isn't chronic Lyme disease after all. Perhaps this is something else. He encouraged her to return to her primary care physician. Perhaps he could help her. Davidson only took care of patients with Lyme disease. He had done all he could.

Discouraged and depressed, she agreed. Her internist referred her to a new rheumatologist and finally, nearly two years after her symptoms started, Carol Ann walked into the office of Dr Linda Bockenstedt at Yale School of Medicine. As she sat in the dreary waiting room, Carol Ann wondered if she'd made a mistake. This looked more like a clinic than a regular doctor's office. There were nearly two dozen patients waiting for any one of a whole string of doctors whose names she'd seen on the door as she came in. Finally she was taken to a small, brightly lit exam room. There were no pictures on the walls, no personal items on the desk. It was as cool and impersonal as a room in a chain hotel.

That chill left the room the moment Bockenstedt entered. She was tall with light hair and warm brown eyes. After introducing herself, she sat on a metal chair and, looking Carol Ann in the eye, she asked her why she had come. And then she listened – without interrupting – as Carol Ann went through her entire story. She described how she was initially diagnosed with Lyme disease, and the crazy symptoms that followed. She recounted the long unsuccessful treatment with antibiotics and the terrible toll it had taken on her stomach and her body. Now during the day she could barely move her arms, her shoulders hurt, and at night her hips and knees throbbed, making sleep almost impossible. She was tired; she could barely concentrate. Her memory was shot. She felt irritable and lost her temper frequently. Bockenstedt took notes as Carol Ann spoke, and when she was done asked a few more questions to help her parse the puzzling symptoms.

Bockenstedt then examined Carol Ann carefully, paying particular attention to her hurting joints. Her neck and shoulders were tender to the touch and too stiff to move fully. Her hands, the joints in which are most frequently involved in rheumatoid arthritis and lupus, were supple and without pain. During the examination Carol Ann's hips and knees were pain-free and had a normal range of motion but she complained that they hurt at night and that in the morning they were so stiff it was hard to get out of bed. The rest of the exam was unremarkable.

By the end of the visit, Bockenstedt focused on three possible diagnoses. First – and most concerning – was a disease not of the joints but of the blood vessels, a disease known as giant cell arteritis. This disease is most common in women over fifty and attacks large blood vessels in the body. Untreated, it can cause blindness and strokes. The most common symptoms are fatigue, weight loss, and body aches – which Carol Ann had – as well as headaches and jaw pain – which she didn't have. Still, it wasn't a disease that Bockenstedt wanted to risk missing.

Another possibility, and the one Bockenstedt thought most likely, was a common but poorly understood disease of the muscles and joints known as polymyalgia rheumatica, abbreviated as PMR. This disease frequently causes stiffness of the neck, shoulders, and hip joints as well as fatigue and sometimes fevers. One of the most interesting aspects of PMR is that it appears full blown out of the blue. Patients often say that they feel like they came down with a flulike illness that never went away.

Rheumatoid arthritis was the third possibility – her symptoms weren't classic, but if not treated, this type of arthritis can cause permanent injury to the joints.

Bockenstedt explained her thinking to Carol Ann and sent her to the lab to look for evidence of any of these joint diseases and to test again for Lyme disease. She also ordered X-rays of her shoulders, which would reveal evidence of rheumatoid arthritis joint damage if it existed.

Two weeks later, Carol Ann was again sitting in Bockenstedt's exam room. Bockenstedt didn't waste any time: she was very confident that Carol Ann had polymyalgia rheumatica. The X-rays ruled out rheumatoid arthritis and the blood work showed no signs of a bacterial infection – by the Lyme bacteria or any other type of bacteria.

Somewhat ironically, there is no test for polymyalgia rheumatica. Instead, tests are used to rule out other possible candidates, and a diagnosis is based on these tests and the patient's constellation of symptoms. Bockenstedt explained why she felt the evidence for PMR was compelling. Carol Ann had a classic presentation of the disease. To begin with, women are four times more likely to get PMR than men. Carol Ann was older than fifty – the age group at highest risk of the disease (one of every two hundred women over fifty get PMR). Her symptoms came on suddenly and felt like an infection. Her pain was located primarily in the large supporting joints of the body – the shoulder and neck, hips, and knees. The spine and the smaller joints of the arms and feet are typically spared. And, of course, the X-rays and blood tests showed no signs of other rheumatologic disease or infection.

Carol Ann took in all of the information without speaking. If this was really true, then Dr Davidson had been wrong and she had endured those months of nausea needlessly. She wasn't quite ready to trust this new diagnosis, even though she trusted Bockenstedt. She had also trusted Davidson. Plus she had read on the Internet that prednisone – the medication that Bockenstedt was suggesting – could worsen a hidden infection, if she had one.

'So you really don't think I have chronic Lyme disease?' Carol Ann asked.

Bockenstedt paused.

This was dangerous territory. Bockenstedt knew from bitter firsthand experience that the 'Lyme literate' – whether physicians or patients – could be ferocious in their attacks on doctors who were sceptical about this disease. She had trained at Yale and watched in 2000 as the 'literate' protested outside Allen Steere's lab. They shouted and carried signs accusing the onetime hero of Lyme disease of being a killer and a monster. He had been heckled and had even received death threats. Why? Because he agreed with the data. He stated publicly that no evidence supported the administration of repeated courses of antibiotics following an initial treatment for Lyme disease. And just last year when the Infectious Diseases Society of America had come out against the treatment of Lyme with months of antibiotics, ILADS charged the members with being in the pockets of the insurance companies who didn't care about the patients and simply didn't want to pay. (Bockenstedt's vivid memories of the ad hominem attacks even made her reluctant to participate in my writing of this book, though in the end her commitment to educating the public won out.)

Bockenstedt therefore weighed her words to Carol Ann very carefully.

There was no evidence that Carol Ann had Lyme disease right now, she told her. None of the tests she had were positive by CDC criteria. In reviewing DeVries's records, Bockenstedt noted that in the first two tests, done by Dr Davidson, a couple of bands in the Western blot had been positive but not the five recommended by the CDC. And in the tests Bockenstedt had run, none of the bands had 'lit up.' Carol Ann's symptoms and the weight of the tests all convincingly pointed to polymyalgia rheumatica as the source of her ongoing symptoms.

'No,' she told Carol Ann. 'I don't think you have chronic Lyme disease. You've been through a lot, I know. But I think you can trust this diagnosis.'

Any doubts Carol Ann might have harboured vanished within days of starting the treatment. The prednisone worked quickly and soon her joints were pain-free. After two years of insomnia, she finally slept through the night. Her flulike feelings disappeared. She could think, she could concentrate, she could remember. She felt like a new woman.

That was four years ago. Carol Ann stayed on prednisone for just over a year, tapering her dose slowly at the end as Bockenstedt recommended to let her body adjust. Since then she's had one flare-up of the old symptoms, but a week or so of prednisone tamed the pain and eased the stiffness.

So, did Carol Ann *ever* have Lyme disease? Probably not, Bockenstedt told me, but it's impossible to know for sure. Certainly by the time she'd come to Bockenstedt's office she had no evidence of the disease. Her painful joints weren't swollen – as they usually are in a Lyme-related arthritis. And none of the test results that Carol Ann had had for Lyme reached the level of confidence required by the CDC guidelines. It could be that Carol Ann was one of those in whom the Lyme bacteria were killed before her antibody defenses formed. Or maybe the rash that the ER doctor based his diagnosis on was a single hive, left over from her attack earlier in the week. Bockenstedt strongly suspects that, in fact, Carol Ann had suffered from PMR all along, but she is careful to acknowledge she can't know for sure. Which is exactly how real diagnosis of real disease often works.

We have tools that are essential in the making of a diagnosis. You have the history. You have the physical exam. You have tests. Ultimately you have treatment. All pieces of the puzzle, all clues leading to the final diagnosis. Says Bockenstedt, 'If I had treated Carol Ann with steroids and she hadn't gotten better, I'd have to wonder – is this really what she has?' But the steroids did work – nearly perfectly. And so while Carol Ann had a good story for Lyme disease, her physical exam wasn't consistent with that disease, her testing wasn't consistent with that disease, and the treatment didn't help her. 'Seeing all this, I don't understand how anyone could argue that Lyme disease is what was causing her all that pain,' Dr Bockenstedt concludes.

Testing has changed how medicine is practised. Doctors can now be far more certain of a given diagnosis with the help of tests than ever before in the long history of medicine. But tests don't make a diagnosis – thinking does. Better tests certainly make for better thinking and Lyme disease sure could use a better test. Until there is one, and possibly long afterward, concerns about missed diagnoses of Lyme and late diagnoses of Lyme will continue to be a source of confusion and contention for doctors and patients alike.

# PART FOUR

## *Limits of the Medical Mind*

## CHAPTER NINE

## *Sick Thinking*

David Powell sat quietly in the tiny emergency room cubicle. His muscular arms and chest were barely covered by the thin cotton hospital gown. He looked far too robust to be in the hospital, and yet this was his fourth emergency room visit in two months. 'I'm losing my strength,' he explained quietly to Dr Christine Twining, a young physician-in-training. 'Doctors keep telling me I'm not having a heart attack. Okay, that's good, I'm glad it's not my heart. But can't anyone tell me what *is* wrong?'

It had started a couple of months earlier, when twenty-seven-year-old David noticed that his hands and fingers felt numb. Then he started having chest pains – a strange tightness or heaviness that made it hard for him to breathe. That's what sent David to his local emergency room the first couple of times. His mother had recently died of a heart attack and he was afraid he was having one too. Once the ER doctors heard his story, they too thought it was his heart. But at each visit the EKG was normal, the blood tests showed he wasn't having a heart attack, and the stress test suggested he wasn't likely to have one anytime soon. That was reassuring, but it wasn't an answer.

As autumn turned to winter, David began to have a hard time keeping up at work – quite literally. He was a garbage collector and noticed that the short sprints from house to truck, which had been part of his daily routine, now left him panting. And the cans he emptied were somehow heavier. His muscles hurt constantly; he had frequent cramps. By the end of the route, his arms and legs shook with fatigue.

'The guys would ask me what's wrong, 'cause I'm strong, a weightlifter, and I was too ashamed to tell them I was getting weak,' David told the young doctor. 'I'd just tell them I'd worked out at the gym real hard the day before.'

The truth was, he hadn't been able to work out for weeks. He simply didn't have the strength. There were other symptoms too: he was losing weight – twenty pounds in two months. And he was tired. After work, he'd nap, get up for supper, then go back to bed. He also had terrible constipation.

Then, just before Christmas, he was shopping with his wife and kept bumping into the shoppers crowding the mall. 'I couldn't make myself go straight,' he said. And his chest felt squeezed, as if he were wearing some kind of girdle around his rib cage. When he began to stagger that evening his wife insisted he go to the emergency room once more. During this visit, his third, there was another normal EKG, another normal set of blood tests, another doctor reassuring him and his wife that he wasn't having a heart attack.

A week later, he almost fell off the back of his truck. 'My fingers were so weak,' he said, 'I couldn't hold on tight. Just one bump and I would have been on the ground.'

That's what brought David to the ER this time. As he told his story – his voice soft and level – he studied his hands, describing their disobedience: these days he had to use both hands to hold up his coffee cup; his handwriting had become a childish scrawl – barely legible even to himself; his fingers could no longer distinguish the coarse cotton of his work clothes from the smooth silkiness of his Sunday tie.

When David returned to the Emergency Department this time, the ER doctor had again gone for the EKG and the blood tests to look for evidence of a heart attack. It's practically a reflex when someone presents with a chief complaint of chest pain. Still, as he reviewed the young man's chart the doctor knew that these were unlikely to provide any insight into what brought this guy back time after time. Emergency room physicians are trained to diagnose and treat life-threatening illnesses – true medical emergencies. For the majority of patients who come to an emergency room and do not have these immediate emergencies, ER physicians make another very important decision about their care: does the patient need to be admitted to the hospital or is this something that can be followed up as an outpatient? While this guy had one of the key symptoms we are all taught to attend to – chest pain – the ER physician thought that it was unlikely the usual chest pain workup was going to provide this man with the diagnosis he sought. So he asked Christine Twining, one of the internal medicine residents-in-training, to see him and admit him to the hospital so that someone could figure out what was going on. While it probably wasn't an emergency, it seemed to him that this young guy really was sick.

Twining listened carefully to David's story. He was so young and healthy-looking. What could be wrong with him? He was only twenty-seven years old; he didn't smoke or drink. He lived with his wife and their six-year-old daughter. Although his mother had died at fifty-five of a heart attack, and two cousins had sickle cell anaemia, the rest of his family was fine.

Physically he was massive. He was just over six feet and weighed 240 pounds – lifting weights had chiselled off most of the fat, so what was left was muscle. On examination there was no evidence of atrophy in the well-defined muscles, and while he easily passed the standard doctor's-office tests of strength, Twining thought that was because those tests were not designed for someone with greater-than-average strength, like this young man.

He'd complained of numbness in his hands and feet. When Twining examined them they certainly looked normal, but when she jabbed them with the pointed instrument she used to test sensation, he couldn't feel it. And when she tapped his joints with her small rubber hammer, the usual spontaneous jerk was completely absent. He had no reflexes. When she asked him to close his eyes and tell her whether she had moved his big toe up or down, testing one of our most primitive senses, he couldn't even tell her that.

Then the doctor noticed a result from a blood test done on one of the patient's previous visits to the ER: he had a low red blood cell count.

Anaemia is unusual in an otherwise healthy young man. He had two very different symptoms – anaemia and this odd weakness and loss of sensation. Were they linked? She couldn't know based on the data she had so far.

She focused first on his loss of strength and the loss of sensation in his arms and legs: having both problems made it clear it was his nerves – and not his muscles – that were the problem. There were dozens of possible causes for this type of neuropathy: diabetes, alcohol abuse, syphilis, HIV, thyroid disease, cancer. But none really fit this patient.

Given his occupation, the doctor considered an unusual cause of this type of nerve damage: toxins. Could he have been exposed to some dangerous substance discarded thoughtlessly or illegally in the regular garbage? Arsenic could cause this kind of nerve damage; lead and mercury could as well. Moreover, these toxins could account for the anaemia as well as the neuropathy, if they were in fact linked.

And what about the anaemia? Had his low red blood count predated this new illness? Sickle cell anaemia ran in his family, and although he had no symptoms of this painful disorder, could these chest pains be from this trait? He complained of abdominal pain: could he be losing blood in his stomach or intestines? It was possible, though his stools had not shown evidence of blood when tested.

The report from the lab described a few abnormal white cells in his blood: the cells contained irregularly shaped nuclei. This suggested he was anaemic because of a nutritional deficiency. A diet poor in folate or vitamin $B_{12}$ could cause anaemia as well as this type of abnormal white cell. Moreover, vitamin $B_{12}$ deficiency could cause neurological symptoms too. It's easy to get adequate amounts of vitamin $B_{12}$ in a normal diet, and it seemed very unlikely that this well-nourished young man could have such a deficiency. But the doctor needed to be sure because the body can't make its own vitamin $B_{12}$ and a true deficiency can cause permanent disability – even death. And treatment is easy and safe: replacing the missing vitamins usually reverses all the symptoms.

Twining sent off blood samples to look for the origin of the anaemia and for evidence of a recent exposure to mercury and arsenic. Other causes of this neuropathy, she thought, were much less likely, and she could test for them later, if necessary.

The results of the anaemia workup came back first. David had no evidence of sickle cell disease or any other congenital blood disorders. He had normal levels of iron and folate. But his level of vitamin $B_{12}$ was dangerously low: a tenth of the normal level. The doctor was sure this was the cause of David's weakness, numbness, constipation, and anaemia. It could even account for his chest pain and shortness of breath.

The cause of David's anaemia was proved by yet another blood test. He had an autoimmune disease that goes by one of those great nineteenth-century names: pernicious anaemia. In this disease, the body's own immune system mistakenly destroys the protein responsible for absorbing the vitamin from digested food and getting it into the blood. The immune system makes antibodies for this protein, just as if it were an invading virus or bacteria. David was started immediately on vitamin $B_{12}$ injections – he would have to take $B_{12}$ supplements the rest of his life. The results were dramatic and almost immediate.

'Every day I can feel myself growing stronger,' David told me when I called him not long after his diagnosis. One week after his first injection he was able to go back to work. 'I can finally run again. I can pick up my daughter again. I can tell I'm going to get it all back.'

## *When Thinking Goes Awry*

David's story is an example of a diagnostic error. Researchers define diagnostic error as a diagnosis that is wrong, missed, or delayed. And although Powell didn't suffer any permanent harm and has been restored to full health, it took four visits to the emergency room to get there.

David was lucky. Many studies show that diagnostic errors often exact a tragic toll. Diagnostic errors are the second leading cause for malpractice suits against hospitals. And a recent study of autopsy findings identified diagnostic discrepancies – a difference between the diagnosis given in life and that discovered after death – in fully 20 per cent of cases. The authors of that study estimate that in almost half of these cases knowledge of the correct diagnosis would have changed the treatment plan. Extrapolated to the millions of people in the United States alone who receive medical care every year, that 10 per cent of diagnostic errors means a vast toll of avoidable suffering and death.

And patients are worried. One survey showed that over one third of patients surveyed after visiting an emergency room had concerns about medical errors and by far the greatest concern was the possibility that they had been misdiagnosed. They are right to worry. A recent review of the data reported that primary care physicians – those in family practice and internal medicine – had a diagnostic error rate that ranged from 2 to 10 per cent. Up to one in ten patients seen was incorrectly diagnosed.

Of course that number only looks at single visits, and anyone who has been to the doctor for a complicated problem knows that it is often figured out over the course of several visits. Emergency room doctors have a somewhat higher rate of diagnostic errors, specialists a somewhat lower rate. This doesn't mean that specialists are better doctors or emergency room physicians are worse. The uncertainty surrounding a diagnosis and thus the likelihood of error is greatest when a patient first presents with a problem – in an emergency room or a primary care office. By the time patients get to specialists much of the uncertainty about their diagnosis has been resolved.

There are many ways of getting a diagnosis wrong. In earlier chapters I looked at how each element of the medical data gathering can break down and lead to diagnostic mistakes – taking an inadequate history or performing an ineffective exam, or not examining the patient at all. A misreading or misinterpretation of a test can also derail the diagnostic process. But perhaps the most common type of diagnostic error – and the one that I will focus on in this chapter – is the one that takes place in the doctor's head: the cognitive error, what I call in this chapter sick thinking. (Anyone interested in learning even more about this important issue should check out Jerome Groopman's outstanding book on the topic, *How Doctors Think*.)

So how often is an error due to sick thinking? Mark Graber, a physician and researcher at the VA Hospital on New York's Long Island, wanted to answer that question. He collected one hundred cases of medical error from five hospitals over the course of five years. For each case, records were reviewed and, when possible, the doctors involved were interviewed within one month of the discovery of the error. These were serious errors. In 90 per cent of the cases patients were harmed by the error; thirty-three patients died.

Graber divided the missed or delayed diagnoses into three categories. (The three categories overlapped somewhat; not surprisingly, most diagnostic errors were due to multiple factors.) 'No-fault errors' are mistakes that happen because of factors beyond the control of the doctor making the diagnosis. When a disease presents in an unusual and uncharacteristic fashion – as when an elderly person with appendicitis has a fever but no abdominal pain – or when a patient provides incorrect information – as a patient with Munchausen syndrome might do – a diagnosis can be unavoidably missed or delayed. This was by far the smallest category of diagnostic error, present in only seven of the one hundred cases.

Graber found that our complex and often poorly coordinated medical system also contributes to diagnostic error. If a test result was not reported in a timely manner or if there were equipment failures or problems, he assigned the resulting diagnostic mistakes to the category of 'system-related errors.' For example, a urinary tract infection might be missed because a urine sample was left too long before being cultured. Or a pneumonia might be missed because an overburdened radiology department hadn't read a critical X-ray correctly. These were relatively common; more than two thirds of the errors Graber studied involved some component of system failure.

The issues that Graber was most interested in were what he called 'cognitive errors,' by which he meant all errors due to the doctor herself. In his study, Graber attributed more than a quarter of all mistakes made, twenty-eight out of one hundred, to those made due to cognitive errors alone. Half of all the errors made were due to a combination of bad systems and sick thinking.

Graber broke his category of cognitive error down further. Which aspect of cognition was at fault? Was it lack of physician knowledge? Not most of the time. Faulty knowledge was the key factor in only a few of the missed diagnoses, each of which involved a rare condition. Faulty data gathering – an inadequate history, missed findings on the physical exam, or misinterpreted test results – was a more common problem, playing a role in 14 per cent of the diagnostic errors. Faulty synthesis – difficulty putting the collected data and knowledge all together – by comparison, played a role in well over half of the incorrect or delayed diagnoses.

In David Powell's case, both the system and the doctors played a role. Early on in his illness David went to two different emergency rooms. Getting records from one ER to another can be a time-consuming affair. Often emergency physicians don't even try to get them because the chances of obtaining them in time to help the patient are so small. So because David went to a different emergency room, his second visit was a virtual rerun of the first visit. And although the patient told the doctor who saw him at his second ER visit that he'd already been 'ruled out' for a heart attack or myocardial infarction (MI), without the records to confirm it, the ER doctor repeated the studies rather than risk missing this important diagnosis.

Because the records were not available, David's diagnosis was delayed. Graber would define this as a system-related error. Certainly, in an ideal world, a patient's records should be readily available.

But the emergency room doctors were guilty of thinking errors as well. Each found that the patient was not having a heart attack but none, save the last, carried that train of thought to its next logical destination. None of them asked that most fundamental question in diagnosis: *what else could this be?* And because they didn't, the diagnosis was missed.

They might have missed the diagnosis even had they asked such a question. The differential diagnosis for chest pain is long, and while this is a well-described symptom of pernicious anaemia, the disease itself is relatively unusual. But they didn't even try.

In medicine it seems that almost nothing that comes after the words 'chest pain' is even heard. And if you are an adult male with chest pain, the odds are almost overwhelming that you are going to end up with a ticket on what I have heard called 'the MI express.' Far too often those words trigger the cascade of EKGs, blood tests, and even exercise stress tests in search of a heart attack – despite the presence of other signs and symptoms or workups that might suggest a different diagnosis.

Each of these doctors exhibited 'premature closure' – one of the most common types of diagnostic cognitive errors. Premature closure is when a doctor latches on to a diagnosis and 'closes off' thinking about possible alternative diagnoses *before* gathering all the data that would justify going down a particular diagnostic path. In David's case, the doctors' thinking was skewed by two factors: the fact that cardiac problems are so common in the ER, and the potentially dire consequences of a heart attack (which lends urgency and pressure to the task of diagnosis). The doctors heard David describe the classic symptom of a myocardial infarction – squeezing or pressurelike pain in the chest associated with shortness of breath – and began ordering tests and exams aimed at clarifying the suspected cardiac situation. In premature closure, 'Thinking stops when a diagnosis is made.' The symptoms of weakness and numbness were noted in the chart in each of the visits but weren't considered on their own even though they are not part of the typical chest pain presentation. When the 'MI express' pulls out of the station, far too often everything that doesn't fit – like David's complaint about his loss of strength – is left behind.

Pat Croskerry is an emergency room physician and a doctor who has written extensively about diagnostic thinking. The brain, says Croskerry, uses two basic strategies in working to figure things out. One is what Croskerry calls an intuitive approach. This 'nonanalytic' approach works by pattern recognition. He describes it as a 'process of matching [a] new situation to one of many exemplars in your memory which are retrievable rapidly and effortlessly. As a consequence, it may require no more mental effort for a clinician to recognise that the current patient is having a heart attack than it is for a child to recognise that a four-legged beast is a dog.'

This is the instant recognition of the true expert described by Malcolm Gladwell in his book *Blink* – fast, associative, inductive. It represents 'the power of thin slicing . . . making sense of situations based on the thinnest slice of experience.' Intuition leads to a diagnostic mode that is dominated by heuristics – mental shortcuts, maxims, and rules of thumb. This is the diagnostic mode used by the emergency room doctors during David Powell's first few visits to the emergency room with his chest pain and strange weakness.

Croskerry contrasts this almost instantaneous intuitive diagnostic thinking with a slower, more deductive approach to diagnostic thought. As described by Croskerry, this analytical approach is linear. It is a process that follows rules and uses logic to think a problem out. It's the Sherlock Holmes model of diagnostic thought.

Croskerry believes that the best diagnostic thought incorporates both modes, with the intuitive mode allowing experienced physicians to recognise the pattern of an illness – the illness script – and the analytic mode addressing the essential question in diagnosis – what else could this be? – and providing the tools and structures that lead to other possible answers.

For Christine Twining, the doctor who finally diagnosed David Powell with pernicious anaemia, there was no *Blink*-like moment of pattern recognition and epiphany when she first heard him describe his symptoms. One thing seemed clear: he wasn't having a heart attack. She felt the patient's fear and frustration. 'He was afraid I was going to send him home with reassurances that it wasn't his heart and without figuring out what it was. But I couldn't send him home; I didn't have a clue what he had.'

Because there was no instantaneous sense of recognition triggered by Powell's odd combination of chest pain, weakness, and anaemia, Twining was forced to approach the problem systematically, considering the possible diagnoses for each of his very different symptoms and pursuing a slower, more rational approach to the patient that ultimately brought her the answer.

Both types of thinking are essential in medicine. Which to use will depend on the perceived degree of uncertainty surrounding a set of circumstances. The more certainty there is in any given set, the more closely it aligns with some recognised or remembered disease state, the more likely you are to use the intuitive response. The cognitive continuum of decision making, says Croskerry, runs from informal/intuitive at one end to calculated/analytical at the other, and the nature of the tasks runs from quite simple to complex. 'The trick lies in matching the appropriate cognitive activity to the particular task.'

Much of the research that has been done on cognitive errors focuses on the misinterpretation of medical information. In David's case, the doctors who missed his diagnosis of pernicious anaemia focused on only a couple of his symptoms, ignoring the history of numbness and weakness, the abnormalities in his physical exam, even the anaemia, in their concern not to miss a heart attack. But errors can also arise from interpretations of data we're not even aware we are making, thanks to assumptions and biases that we bring with us from our lives outside the hospital.

## *Physician Bias, Fair and Otherwise*

'Doc, my knee, it's doing this thing again.' Vera Freeman pointed to her red and swollen knee as I entered the small, poorly lit hospital room. She was an attractive young woman, with hair stylishly braided and ornamented with bright beads. 'Last night it was just fine,' she reported. 'Now just look at it.'

Two weeks earlier, she had awoken to find her ankle – not her knee – swollen and painful. She didn't remember injuring it. 'It just blew up,' she said, and when she took it easy for a couple of days, it got better. 'But just as soon as it was okay, my wrist swelled up. It was big, and it really hurt. I was getting worried, but it got better too.' The next week, though, her knee began to swell, and she decided to come to the hospital. 'It was so weird. It was like I had this swelling that just didn't know where to light.' She looked at me carefully, to see if I was following her story.

She stayed in the hospital for a couple of days, received some intravenous antibiotics, and was sent home with antibiotics to finish by mouth. She took the pills for a couple of days, but once she felt better, she neglected to take the rest. Now the pain and swelling had wandered back, and she wanted to know why.

Freeman was frank about her history. She had HIV, which had been diagnosed three years before. Otherwise, she thought she was pretty healthy. She did not smoke cigarettes or drink, though she did admit to smoking crack cocaine 'occasionally.' She had no children, lived in an apartment with her longtime boyfriend. She had, at times, worked as a prostitute to help her buy crack.

On exam, her dark brown skin felt warm. Moving the joint elicited a sharp cry of pain. As I gently explored the swollen knee, I could feel fluid moving around, like a warm, firm water balloon. The kneecap was separated from the joint it normally covers; I could press it down almost an inch before I felt contact. As I examined her, I assembled a differential diagnosis in my head. A hot, swollen joint is routine in medicine, usually caused by trauma, by gout, by infection. But this 'wandering' pain was far from routine. In the textbooks it's known as a 'migratory polyarticular arthritis' – that is, an arthritis that moves from joint to joint – and it is an extraordinary manifestation of just a few pretty ordinary diseases.

It is seen most commonly with gonorrhoea (although it is unusual even in this disease), where it is often accompanied by fever and a rash. Lyme disease can also manifest this way, as can viruses like hepatitis and even HIV. But none of those seemed to fit. There were other, less likely possibilities. Rheumatoid arthritis can come on like this, as can lupus.

Justin Thompson, the intern working with me that month, had admitted Freeman for her earlier hospitalisation. When I asked him about her, he wearily flipped through a stack of index cards that he pulled from his pockets. 'Right. We tapped her knee and cultured her up,' he said, meaning that they'd drawn fluid from her knee, which should offer some clues, and had sent off some of the fluid, as well as her blood and urine, to check for evidence of infection. 'I thought it was gonorrhoea,' the intern stated flatly. 'It's not the way you usually see it, but gonorrhoea can definitely cause this.'

The art of diagnosis can look a lot like profiling. Doctors constantly ask: Is a particular condition more common in men or women? Whites or blacks? The young or the old? In this way a doctor narrows the possible causes of a given illness in a given patient. Gonorrhoea, then, was the most likely diagnosis for this young, sexually active, onetime prostitute. And while none of the tests had confirmed it, none ruled it out either.

But here she was again, knee gigantically swollen – again. This was not part of the disease profile, yet the intern working with me was undeterred. So much so that he had already ordered antibiotics to treat her presumed infection. Because she did not finish her course of antibiotics, the disease had been only partly treated; therefore, all she needed was more antibiotics. 'Or maybe her boyfriend was the source,' he said, 'and she's been reexposed since getting treated. Or maybe she's back on the street.'

These were all reasonable thoughts, but it was clear to me that we needed more evidence to make that diagnosis a second time. I thought we should hold the antibiotics until after we tapped the knee again and repeated her cultures.

I was also interested in the results of the blood tests from Freeman's earlier hospitalisation. I found a computer and tracked down her test results. Lyme was negative; hepatitis, negative; gonorrhoea and syphilis, negative. In fact, there was only one set of positive results: the tests for recent strep infection along with several other blood tests consistent with a diagnosis of rheumatic fever. The problem is that rheumatic fever rarely occurs in these days of antibiotics, and when it does it is seen almost exclusively in children. It is practically unheard of for an adult to develop it. Even now that she met some of the criteria for the disease, such a diagnosis was hard to make. She simply didn't fit the profile.

We went back to the patient. Had she had a sore throat recently? Yes. She'd had a sore throat a few weeks before, but she thought it could have been because of the crack. That convinced me. It now seemed clear that, as unlikely as it might have seemed initially, this young woman had rheumatic fever.

When we went back to the patient, she was dressed and ready to leave. Her knee, which had been red and hot and excruciatingly painful only twenty-four hours earlier, had improved significantly with no intervention. We scheduled her to see her doctor the following week. As she stood with her bag in hand, I tried to explain rheumatic fever and what it might mean to her, but she wasn't listening.

'I'm better,' she announced, 'so I'm gone.' I gave her her prescriptions and shook her hand, then watched as she limped down the hall, waved gaily from the door, and disappeared.

Recently I caught up with her doctor, who told me that Vera had had an echo to look for any signs of damage to her heart or their precious valves that direct the flow of blood through the organ. Everything was completely normal. And it made sense. Cardiac injury is very common in the children who get rheumatic fever; in adults the disease tends to 'bite the joints and lick the heart,' causing joint pain but not the more significant cardiac lesions.

What has always stuck in my mind is the intern's insistence on the diagnosis of gonorrhoea even in the face of failed tests for that condition. Was he just being prejudiced against a minority woman with a history of behaviours not sanctioned by the larger society? Possibly, but I think the story is more complicated than that.

At first glance, patients might think that the ideal in diagnosis would be for a doctor to treat (and view) all of their patients identically – to be colour, age, gender, and socioeconomically blind. We don't want our looks to influence our doctor's objective assessment of our health problems. And yet they must. Illnesses and diseases do not abide by our constitutionally required equal protection. Diseases *do* discriminate on the basis of race, gender, age, and even socioeconomic status.

To take a obvious example: the vast majority of breast cancer patients are women, so it is not wrong for a doctor to automatically drop that diagnosis down in her priority list when confronted with a male patient with a lump on his chest. A less obvious example is prostate cancer: black men are significantly more likely to get this type of cancer than men of other races – four times more likely, in fact, than Korean men, nearly twice as likely as men of European descent. So if a black man comes to a doctor complaining of urinary symptoms, a good doctor will automatically raise her level of suspicion for prostate cancer based solely on the colour of the patient's skin.

In fact, it would be irresponsible of the doctor *not* to take race into account when considering this diagnosis.

Viewed in this light, the bias of the intern clinging to his suspicion of gonorrhoea in a woman with a history of drug use and prostitution is not so egregious. Using drugs and having multiple sex partners, after all, are legitimately associated with an increased risk for sexually transmitted infections. What would be egregious is if the intern (or anybody else) settled on a diagnosis of gonorrhoea based *only* on the colour of the woman's skin, her clothing, or some other aspect of her appearance or behaviour, all of which have nothing to do with one's risk of gonorrhoea.

In other words, patients want doctors to be *legitimately biased* in their thinking and decision-making processes when struggling to find a diagnosis. Doctors should take into account any known associations that might help pin down the cause of an illness. But diagnoses can be missed when doctors apply false generalisations or close off diagnostic possibilities just because they are less likely in a certain group or population (e.g., 'This can't be HIV because the patient is elderly'). Research has shown that medical decision making is shaped by many of the same influences that distort other aspects of human interaction. Indeed, says one group of researchers, 'despite their "objective" medical training, physicians remain human actors, socially conditioned to engage in stereotyping, whether consciously or not.' In that respect, medical decision making can be a function of who the patient is as much as what the patient has.

Studies in social science have documented many nonmedical factors that influence medical decisions, including characteristics of the patient such as age, gender, socioeconomic status, race, or ethnicity. These can be important considerations in prioritising possible diagnoses. But characteristics that have no obvious medical meaning – such as the presence or kind of health insurance, assertive personality type, or even physical attractiveness – have also been shown to play a role in how doctors make decisions about medical diagnosis and care. And even those factors that can affect the probability of disease in some cases, factors such as age and sex, are extraneous in many others.

One of the many careful experiments designed to tease out such influences illustrates this point. A set of videotaped doctor-patient encounters was created using professional actors. Scripts were created for male 'patients' and female 'patients' who complained of a set of cardiac symptoms. The scripts and all presenting details were identical in every respect aside from trivial changes in personal pronouns and the like. Two hundred and fifty-six doctors practising in both the United States and the United Kingdom were recruited for the study. They viewed either one or the other video scenario and were then asked a series of questions about what disorder they suspected, what treatments or recommendations they would suggest, and so on. Coronary heart disease (CHD) was chosen because it is the leading killer of both men and women, and although age-specific mortality rates are higher for men than women, twice as many women as men aged forty-five to sixty-four have undetected or 'silent' CHD, which suggests that the true incidence between men and woman may be similar. This is a case, in other words, in which doctors should *not* apply a bias in their decision-making processes – here a gender bias.

The study results, however, clearly demonstrated just such a bias. Gender was found to have significant influence on all aspects of doctors' diagnostic strategies; in each case women received less attention than men presenting with CHD symptoms. Doctors would ask men more questions than women (on average 7 and 5.7 questions, respectively) and perform more extensive examinations for men than women (5.1 compared to 4.3 parts of the body or body systems would be examined, respectively). CHD was mentioned as a possible diagnosis for more men than women (95 and 88 per cent, respectively), and doctors had significantly higher certainty of CHD for male than female patients, 57 and 47 per cent, respectively, on a scale of 0 (total uncertainty) to 100 per cent (total certainty).

The study authors concluded: 'Our findings indicate that women presenting with CHD symptoms are disadvantaged in primary care. Doctors provide a less thorough diagnostic search procedure than for men presenting with identical symptoms, and fewer women are given prescriptions appropriate for treating CHD.'

The impact of conscious or unconscious bias on the diagnostic thinking processes of doctors adds to the complexities of the entire doctor-patient experience. The best doctors acknowledge their vulnerabilities and try hard to retrain themselves or monitor themselves and their thinking processes as they move through any given diagnostic challenge.

The last type of cognitive error I want to talk about is what's often called in the cognitive literature diagnostic momentum. This is a kind of medical groupthink in which once a diagnostic label is attached to a patient, it tends to become 'stickier and stickier.' Doctors are taught in medical school that they should not simply accept a diagnosis given to a patient but should reevaluate the data for themselves before accepting or sometimes rejecting this diagnosis. That we should, as former president Ronald Reagan often exhorted (in a very different setting), 'Trust but verify.' Rather than accept a previous diagnosis, doctors are supposed to start fresh by thinking things through for themselves. This, of course, is much easier said than done.

If a doctor is tired or in a hurry, she is far less likely to take the time to review all the test results and other evidence that went into the diagnosis. And even if she does expend the effort to do that, it's difficult to not fall into the same well-defined disease pattern – potentially mistaken or not – that those who have seen the patient already have defined. But that kind of extra effort can sometimes pay off dramatically.

## *The Doctor of Last Resort*

Graciela Moity spoke in a slow, husky vibrato. She sounded weary, discouraged. 'I can remember clear as day when it all began,' she said. 'It was just over a year ago. I woke up and felt like my legs were on fire.'

She was talking to Dr David Podell – the most recent of a train of doctors who had evaluated the woman since that day she awoke in such pain. The three previous doctors couldn't figure out what was going on. Their best guess was scleroderma, a disease caused by the overproduction of one of the connective tissues, collagen. The patient's symptoms weren't a great fit, but sometimes the disease could manifest itself in unusual ways. She was referred to Podell for confirmation of the diagnosis and treatment of this unusual autoimmune disorder.

With years of experience under his belt, Podell knew that when a patient has already been to a slew of specialists before arriving at your door, you need to approach the case with a different mind-set – with different assumptions. You know, for example, that whatever this patient has, it isn't going to be obvious. Maybe it's an unusual disease, known best by specialists – like scleroderma – or perhaps it's an unusual presentation of a more common illness. In any case, it won't be routine. In such situations, he knew, you had to start from scratch even if the patient comes to you with a diagnosis already made. He asked the woman to continue with her story, apologising because he knew she had told and retold it so many times already in the past year.

She said that until that morning a year ago she had always been healthy. But the burning pain in her legs had been so intense that now she could hardly walk. And she had felt weak – especially in her left leg. She went to her regular doctor, but he didn't know what to make of her symptoms and sent her to a neurologist. He examined her, sent off a dozen blood tests, and got a CT scan of her head and spine before sending her back to her internists, still undiagnosed.

Then she developed a cough. It was usually a dry, irritating cough, but occasionally she coughed up blood. Recently she felt out of breath with even slight exertion. This morning, she told Podell, she had to stop and rest during the short walk from the parking lot to his office. Her internist sent her to a pulmonologist because her lungs seemed clearly involved. He got a chest X-ray, then a chest CT, more blood tests, even a biopsy. The chest X-ray proved that her lungs were involved. In the normally black areas of the image over the air-filled lung tissue, there were faint patches of white. The biopsy showed inflammation but nothing more specific. He wasn't sure what this was. He tried her on a variety of antibiotics. Finally he sent her back to her internist, suggesting the possibility of scleroderma.

Eventually her internist sent her to Podell, who is a rheumatologist – a specialist in diseases of connective tissues. Because connective tissues are found throughout the body, complex, multisystem illnesses are the rheumatologists' bread and butter.

The patient was a slender woman with a mass of straight dark hair streaked with grey. Her skin was clear, but her eyes were puffy with fatigue, and she looked older than her fifty-three years. Examining her, Podell found few obvious signs of disease. Despite the cough and breathing problems, her lungs sounded clear. She had some mild weakness in her left hip, but other than that, her joints, skin, and muscles were all normal.

Podell could see why the previous doctors were puzzled. Her symptoms suggested that her illness involved the nervous system and the pulmonary system, which is an unusual pairing. Although scleroderma can affect both nerve and lung tissue, Moity didn't have the classic thickening of the skin that is the hallmark of that disease. Could this be an atypical form of scleroderma? Or was it something else altogether?

Could this be Sjögren's syndrome, a disease in which the immune system mistakenly attacks a patient's fluid-producing glands? Sjögren's can affect the lungs and sometimes spreads to the nervous system. Patients with Sjögren's usually complain of painful eyes or a dry mouth, and this patient had mentioned that her mouth was dry.

Podell ordered blood tests to look for evidence of Sjögren's. He assured the patient that he would do all he could to figure out what was going on, but that it would take a bit more time. Looking defeated, Graciela Moity made an appointment to return in a couple of weeks and trudged out toward the parking lot.

Podell wanted to examine the patient's extensive medical record, especially the tests and results obtained by the other doctors. He didn't read the records in advance in complicated cases. He felt it was important to take in the information without any preconceived notions about what was going on. But at the end of the day Podell sat down with her thick chart and went through every page. When you are the last in a string of practitioners, one of your most important jobs is to review each piece of the puzzle with fresh eyes, questioning every assumption and double-checking the reported results. In complex cases like this, the answer is sometimes already there, just waiting to be noticed.

A long list of blood tests had been done. Several suggested an inflammatory process, but none identified the cause. The patient also had MRIs of her head and spine, as well as a CT scan of the chest. Podell was particularly interested in the chest CT, which showed something he could not have detected in his physical exam: faint, cloudy patches throughout both lungs. He wasn't an expert in interpreting CT scans, so he called in a radiologist to look them over. But the colleague merely confirmed what Podell could already see: cloudy areas showed the presence of fluid in both lungs. Etiology: unknown.

The patient had also had a lung biopsy. The pathology report said there was evidence of inflammation but, like the blood tests, revealed nothing of the cause. But, again, Podell sought an expert opinion – in this case from the pathologist Tom Anderson. Podell and Anderson sat at a double-headed microscope in the pathology lab, scanning slides that held the biopsy samples. The first slide showed evidence of extensive inflammation, Anderson agreed, but nothing more. As he zipped across the second slide, Anderson reported that again he saw lots of inflammation. Suddenly he stopped. He quickly flipped the microscope lens to zoom in tight on one group of cells that formed a cluster, quite different in appearance from the cells around it.

'That looks like a granuloma,' he said.

These distinctive cell formations are characterised by groups of giant cells up to a hundred times larger than normal cells. They are seen in the lungs only in a few diseases – most commonly sarcoidosis (known more commonly as sarcoid) and tuberculosis. Podell almost laughed out loud. At last, the needle had fallen from the haystack. He picked up the phone and called the patient.

'I know what's going on,' he told her. 'I can explain everything.'

The culprit, Podell explained, was almost certainly sarcoid, a mysterious chronic disease characterised by inflammation of tissues that often display the unusual granuloma collections of cells. The disease usually affects the lungs, but in one third of cases can attack other parts of the body as well, including (rarely) the nervous system. He told her she would need to be tested for tuberculosis, as that disease can also cause granulomas, but he was confident that's not what she had. She had none of the common symptoms of TB such as night sweats, weight loss, or fever. No, Podell said, this is overwhelmingly likely to be sarcoid.

Podell started the patient on the corticosteroid prednisone, which is a highly effective anti-inflammatory medicine. Almost immediately her breathing became easier and the cough disappeared. Within a few days she was walking up and down stairs, something she hadn't been able to do for more than a year. The damage to the nerves in her legs would take longer to treat and may not be completely reversible, but with the diagnosis now clear and effective treatments known, the prognosis for a full recovery was excellent.

Dr Podell wasn't born an excellent diagnostician. He didn't always know to check and double-check the work of other doctors earlier in the 'train' for any particular patient. He learned this and many other invaluable lessons about diagnosis over the course of a long career. And that, in the end, is why we can be hopeful that doctors and other health care providers can avoid or even eliminate the types of cognitive errors we have encountered in this chapter. Yes, doctors are human beings and, thus, are prone to biases, distortions of perspective, and blind spots. But doctors have the capacity to learn from their mistakes, overcome built-in biases, and guard against the kinds of thinking errors that in other professions might only be an annoyance.

I recall a rather mortifying moment in my own learning curve. I was in my third year of medical school. I was given a very simple task by an experienced doctor: to intubate an unconscious patient. Intubation is to medicine what boiling water is to cooking – one of the most basic techniques you can think of. And yet I blew it. Because both the trachea (the tube for air) and the oesophagus (the tube for food) diverge at the back of the throat, it is relatively easy to slide the breathing tube into the oesophagus. Doing so, of course, is a potentially deadly mistake. Students are therefore repeatedly taught to listen to the lungs for sounds of air movement after placing the breathing tube. If you've accidentally put the tube into the stomach, the lungs will be silent. When I listened I heard the terrible silence that means you've made this basic error. Under the gaze of my supervising doctor, I removed the tube and tried again, feeling extremely embarrassed in the process. But the doctor was not annoyed or disappointed. And what he said next has always stuck with me.

'There's no shame in intubating the oesophagus,' he said. 'But there *is* shame in not checking or catching the error.'

His point was that errors themselves are unavoidable. Mistakes will always happen – all types of mistakes, from the technical to the cognitive. But that doesn't mean we throw up our hands in helplessness. The key is designing our systems, our procedures, our protocols, and our own thinking processes to minimise mistakes as much as possible and then to *catch* mistakes when they are made.

Medicine is not the only field in which mistakes can be deadly. The airline industry, to take just one example, has had to put into place many systems for preventing and catching human errors. In the 1930s, following a crash in which a test pilot and crewman were killed due to 'pilot error,' the air force responded by requiring every pilot and copilot to complete a pretakeoff checklist before each flight. The rate of accidents plummeted, and eventually this became standard practice for military and commercial pilots. Most airlines also now require pilots and crew to review the flight plan just before takeoff. This is done as a group and anyone in the crew, from pilot to steward, can bring up any problems they see or anticipate. Pilot and crew are drilled on safety procedures for a wide variety of problems, often using flight simulators to make the experience as real and useful as possible. These basic steps are part of a broader movement that has dramatically improved air travel safety.

There is a national effort now being made to eliminate many of the errors in medicine, to implement layers of checks and double checks to catch errors before they happen. Many of the strategies developed by the airline industry have been adapted and incorporated into hospitals and operating rooms throughout the United States. For example, there is an effort to require surgeons to complete a pre-surgical checklist with all the members of the surgical team. Before any operation, the team meets and anyone, from the anaesthesiologist to the scrub nurse, can bring up any problem they see or anticipate. A recent study in the *New England Journal of Medicine* showed that the use of a nineteen-item surgical safety checklist decreased mortality by nearly 50 per cent and the rate of complications overall by a third. A recent study showed that the use of a checklist before certain procedures in the ICU can also reduce medical errors by 80 per cent and save lives.

Most of this effort has been directed at system errors – when the wrong drug is given or the wrong type of blood transfused. When the wrong leg is amputated. These were the errors identified in a report by the Institute of Medicine (IOM), *To Err Is Human,* published in 2000. Hospitals have been in the forefront of this movement and there are efforts to punish hospitals that have been slow to address these problems.

Diagnostic error, however, hasn't been part of that effort. In fact, when one researcher searched the text of the IOM report, the term 'medication error' came up seventy times but the term 'diagnostic error' came up only twice. This was true even though the study that this report was based on found that diagnostic errors accounted for 17 per cent of all the errors made.

Research into the cause and solution of diagnostic error is in its infancy. Most of the focus in the area of diagnostic errors is aimed at addressing one of the most fundamental cognitive limitations that doctors must deal with: the limited capacity of our own brains. Medical knowledge has grown so vast that no single human can know it all – no matter how much experience they have, no matter how many patients they've seen, and no matter how many textbooks they've read, no matter how many journals they keep up with. Some classes of cognitive errors are rooted in this limitation – you can't see what you don't know to look for. And even if you know about an illness, you may not think of it if a patient presents with an unusual version of the disease.

One obvious solution to this dilemma is for doctors to augment their own, personal neural computers with actual computers, which don't get tired, don't get confused, and have memory capacities that far outstrip that of any single human brain. But, as we will see, this 'obvious' solution has not been nearly as easy to implement as many medical professionals once believed.

# CHAPTER TEN

## *Digital Diagnosis*

In 1976, Peter Szolovits had a vision of the future. He had a newly minted doctorate in information sciences from Caltech. He was in the vanguard of the computer-savvy. And he had a dream: that joining the data-gathering skills of the physician with the almost limitless memory and data-crunching ability of the computer would allow unprecedented accuracy in the physician's art of diagnosis.

Szolovits came of intellectual age in a time of heady optimism about the capacity of these marvelous inventions. It was the dawn of the modern computer era. Microcomputers were the cutting edge. These computers were the size of a desk rather than the room-sized mainframes that had been the previous state of the art. The personal computer – one that ordinary people could use in their homes – was still just a dream in a Palo Alto garage. Data was still stored on enormous reels of electromagnetic tape. The newly invented LP-sized disk drives were marvels of data storage technology because they could hold 7 megabytes of information.

The rapidly growing ability of computers to store vast amounts of information seemed to fit perfectly with the needs of medicine, in particular the challenges of medical diagnosis. It was obvious that medical knowledge was also growing exponentially. In a 1976 article, a group of doctors working on a computer simulation of 'clinical cognition' estimated that a practising doctor draws on a store of at least two million medical facts. And it was clear that this mountain of knowledge would only grow larger with time. Using a computer 'brain' to augment and support human brains in the often bedeviling work of diagnosing illness seemed to Szolovits a logical and technologically feasible goal.

During these heady times Szolovits began a series of conversations with physicians about collaborating to design a computer to help doctors meet the demands of the rapidly expanding universe of medical knowledge. He was surprised by what he found. One conversation in particular with a highly respected senior physician in a university teaching hospital stands out from those days. After listening to Szolovits describe the possibilities of, for instance, entering a set of symptoms into a computer that would then generate a list of likely diagnoses, the physician interrupted him.

'Son,' he said, raising his bare hands in front of Szolovits, 'these are the hands of a surgeon, not a typist.' And he turned on his heel and walked away.

It was an early indication that the application of computers to medical diagnosis might not be as straightforward as Szolovits had thought.

Flash forward thirty years.

By 2006, Szolovits was a full professor at MIT. An energetic man with just the barest hint of middle-aged thickening and a salt-and-pepper beard, Szolovits heads the group at the Massachusetts Institute of Technology devoted to designing computers and systems of artificial intelligence to address problems of medical decision making and diagnosis. Every autumn he shares his ideas and insights into this world in a graduate student seminar called Biomedical Decision Support. I had read about this course and wanted to see what the future of diagnostic software was going to look like.

I visited at the end of the semester, when students presented their final projects. Sitting in a hard plastic chair in the classroom, I watched as PowerPoint slides whizzed by, accompanied by rapid-fire, acronym-studded sentences. One group presented a new technique to look for 'interesting hits' amid vast databases; another presented a user-friendly interface for a Web-based electronic medical records program; a third presented a program that bolsters the privacy of genetic test data. One group exceeded their fifteen-minute slot to describe an elegant program for identifying potentially harmful interactions between prescription drugs that performs better than the current state-of-the-art software.

All of the projects seemed to improve or expand the boundaries of one or another aspect of health care delivery. Indeed, after the presentations Szolovits chatted with the team who'd created the drug interaction program because not only did it appear to be publishable, it might also be something the students could turn into a business opportunity.

And yet something was missing. Despite the title of the course, none of the projects addressed the issue that had beckoned so alluringly to Szolovits thirty years ago – the task of improving clinical diagnosis with computers.

In his office after the class, Szolovits leaned back in his chair, musing.

'Thirty years ago we thought we could identify all of the best practices in medicine, create a system that would make diagnosis faster and easier, and bring it all to doctors via a computer,' he said. Twenty years ago he wrote a paper for the *Annals of Internal Medicine* that proclaimed artificial intelligence techniques would eventually give the computer a major role as an expert consultant to the physician. And today? Szolovits sighed. 'As it turns out, it's simply not possible.' It might be an interesting idea, but there's no market for it. Doctors aren't interested in buying it and so companies aren't interested in designing and building it. 'Rather than trying to bring the average doctors up to a level of being super-diagnosticians, the emphasis and attention has shifted toward bringing below-average doctors up to current standards and helping even good doctors avoid doing really stupid things. That turns out to provide greater benefits to patients. Plus, there is a financial model for it.'

Szolovits ticked off some of the major reasons that most doctors today still rely on their own brains and the brains of their colleagues when making a diagnosis rather than a computerised diagnostic aid.

First, computers can't collect the data from the patients themselves. These machines excel in data crunching, not data collecting. Physicians must collect the data and then enter it into the program. And the programs themselves don't make this easy. There are many ways of describing a patient's symptoms and physical exam findings, but most computers don't have adequate language skills to understand. You're left with long pull-down lists of every possible symptom variation or using terms that the computer simply doesn't recognise.

There are technical difficulties as well. Doctors, laboratories, and hospitals all use different kinds of computer software. No single system can interface with the huge variety of software used to store patient data. Once again the physician must provide the data if she wants it to be considered. Then there are financial difficulties. Who is going to pay the doctor or hospital to invest in this kind of software? Szolovits noted that hospitals don't get reimbursed for *understanding* things, they get reimbursed for *doing* things.

But perhaps the greatest difficulty lies in persuading doctors themselves to use this kind of software. When confronted with a confusing clinical picture, it is often faster and easier for doctors to do what doctors have always done – ask for help from other doctors.

For these and many other reasons, the medical community has yet to embrace any particular computerised diagnostic support system. The dream of a computer system that can 'think' better, faster, and more comprehensively than any human doctor has not been realised. For all their limitations, well-trained human beings are still remarkably good at sizing up a problem, rapidly eliminating irrelevant information, and zeroing in on a 'good-enough' decision.

This is why human chess players held out for so long against computer opponents whose raw computational and memory abilities were many orders of magnitude better than those of a human brain. Humans devise shortcut strategies for making decisions and drawing conclusions that are simply impossible for computers. Humans are also extraordinarily good at pattern recognition – in chess, skilled players are able to size up the entire board at a glance and develop a feel, an intuition, for potential threats or opportunities.

It took decades and millions of dollars to create a computer that was as good as a human at the game of chess. It is a complex game requiring higher order thinking but is two-dimensional and based on clear, fixed rules using pieces that never vary. The diagnosis of human beings, in contrast, is four-dimensional (encompassing the three spatial dimensions and the fourth dimension of time), has *no* invariable rules, and involves 'pieces' (bodies), no two of which are exactly the same.

In addition, of course, humans have a set of diagnostic tools that computers may never equal – five independent and exquisitely powerful sense organs. At a glance, a doctor can take in and almost immediately process reams of information about a patient – their posture, skin tone, quality of eye contact, aroma, voice quality, personal hygiene, and hints and clues so subtle they defy verbal description. A computer, in contrast, has only words and numbers, typed in by a human, that inadequately represent a living, breathing, and immensely complicated patient.

Despite the challenges, Szolovits was among those who first attempted to develop computer programs to diagnose medical conditions. Dozens of prototype models were created and tested in a laboratory setting. But most foundered when attempts were made to scale them up, move them into a clinical setting, or make a profit on them. Computers lacked the necessary memory and processing speeds to make vast databases rapidly usable. Until the advent of the World Wide Web, programs had to be distributed via diskettes, or as part of a dedicated computer, or via dial-up modem connections. All of these challenges slowed momentum in the field.

But even systems that have embraced more recent technological improvements have not seen wide success. A case in point is one of the earlier attempts to use computers to improve diagnosis. In 1984 a team of computer scientists from MIT's Laboratory for Computer Science teamed up with a group of doctors from Massachusetts General Hospital, just across the river. They worked for two years to develop an electronic medical reference system and an aid to diagnosis. In 1986 the program, dubbed DXplain, was launched with a database of information on five hundred diseases. National distribution of DXplain with an expanded database of about two thousand diseases began in 1987 over a precursor to the Internet – a dedicated computer network using dial-up access. Between 1991 and 1996, DXplain was also distributed as a stand-alone version that could be loaded on an individual PC. Since 1996, Internet access to a Web-based version of DXplain has replaced all previous methods of distribution. The program has been continually expanded over the years and is now available to about 35,000 medical personnel, almost all of them in medical schools and teaching hospitals where the program is used as an educational tool.

DXplain and other first-generation diagnostic decision support software programs use compiled knowledge bases of syndromes and diseases with their characteristic symptoms, signs, and laboratory findings. Users enter the data from their own patients by selecting from a menu of choices, and the programs use Bayesian logic or pattern-matching algorithms to suggest diagnostic possibilities.

'There was a lot of work in the 1980s on using computers in diagnostic problem solving and then, in the 1990s, it sort of petered out,' says Eta Berner, a professor of Health Informatics at the University of Alabama. Berner may have been part of the reason this work petered out. In 1994 she and a group of thirteen other physicians tested four of the most widely used programs in a paper published in the *New England Journal of Medicine*. They collected just over one hundred difficult cases from specialists from around the US. They entered the data from each of the patients into each of the four databases. All four programs correctly diagnosed 63 out of the 105 cases included in the study. Overall the four programs provided the correct diagnosis anywhere from 50 to 70 per cent of the time – a solid C performance at best.

The authors of the study concluded that the programs tested might be somewhat helpful in clinical settings: 'The developers of these systems intend these programs to serve a prompting function, reminding physicians of diagnoses they may not have considered or triggering their thinking about related diagnostic possibilities.' But as their study showed, many times the programs would not provide the answers that the doctors are looking for. 'The field was sort of a wasteland for a while,' Berner explained, but then added, 'Now it's picking up again.'

## Consulting an Expert System

One of the difficulties of diagnostic software systems like DXplain is that they try to cover all areas of medicine. Other systems that have been developed as specialised 'expert systems' are used by doctors when a case presents a particular type of diagnostic challenge.

Dr Frank Bia is the medical director of AmeriCares, an international relief organisation. He's also a specialist in infectious disease – particularly tropical disease – and until recently a professor of medicine at Yale. He uses a program called GIDEON (Global Infectious Disease and Epidemiology Network) when he sees patients who are sick and have recently returned from other countries. Not long ago he described a case where GIDEON provided clues to a very difficult diagnosis.

It was the early hours of the morning. A twenty-one-year-old woman was moaning softly in her hospital bed. Beside her an IV dripped fluid into her slender arm. Her mother sat next to the bed, her stylish clothes rumpled from her night-long vigil and her face heavy with fatigue.

She'd been brought to the emergency room of this small Connecticut hospital late one night, pale and feverish. 'She's been like this for two weeks,' the mother told the young physician who entered the room. 'And no one can figure out why.'

Her daughter had always been very healthy. She'd recently spent a month on a research trip to Africa without any health issues. It wasn't until two weeks after her return to Wesleyan College that she began to feel hot and sweaty. Just standing up made her light-headed. A lengthy nap brought some relief but, by the next day, she realised that she was feverish, so she went to the infirmary.

'I told them I thought it might be malaria,' the patient explained to the doctor in a barely audible voice. 'The teacher told us it was common where we were in Tanzania.' And she hadn't always taken the preventative medicine while she was there. The school nurse thought it was probably the flu. But when the young woman didn't get better over the next several days, the nurse referred her to an infectious disease specialist in town. Maybe it *was* malaria. Since she had been in an area rife with this mosquito-borne illness, the specialist started her on a week of quinine and doxycycline.

She took a full seven-day course, but the medicine didn't help. Over the next few days she developed a cough so violent it made her vomit. She had abdominal pain that made even standing difficult. And she had terrible diarrhoea. When she made yet another trip to the infirmary, they called an ambulance to take her to a hospital nearby.

Fadi Hammami, the doctor on duty that morning, listened quietly to the story. He told me later: 'I didn't want to miss this diagnosis. She probably had picked up something in Africa; I just had to figure out what it was.'

Lying on the stretcher, the patient was thin and pale; her skin was stretched tight across her cheekbones. She had a temperature of 102°. Her blood pressure was low, and her heart was beating fast and hard. She had good bowel sounds, and although her belly was tender, he found nothing else out of the ordinary.

He turned to the lab results sent earlier that morning. Her white blood cell count was elevated, indicating an infection. Some of her white cells were enlarged and their nuclei were irregularly shaped. And something else in her blood work intrigued the doctor: nearly half of her white cells were a single type of infection-fighting cell – eosinophils. Normally these make up only 2 to 7 per cent of a person's white cells. In this patient, eosinophils accounted for 41 per cent of the white cells in her system. He'd rarely seen that before, and it was an important clue. This type of cell is the body's most effective defense against one class of infectious agents: parasites.

But which parasite? There are dozens, each with a different treatment. Trichinosis, caused by a tiny worm transmitted through infected meat, was capable of this kind of illness. It is rarely seen in the US but is endemic in many African nations. *Strongyloides,* a parasite that lives in contaminated soil, is also known to cause this type of white cell response, as is filariasis, a disease transmitted by mosquitoes. Which agent was most common in the area of Tanzania she visited?

Dr Hammami knew he needed help. Dr Frank Bia provided it. Dr Hammami had heard of the doctor and called him. He introduced himself and quickly launched into the details of the case. Dr Bia took notes as he listened. He immediately realised that the list of diseases that cause such a profound eosinophilia was short. Trichinosis, he told Dr Hammami, was unlikely because the patient didn't have muscle pain. Filariasis was a much more slowly progressing disease, usually causing symptoms months rather than weeks after the exposure. Strongyloidiasis was a good possibility. So was another disease, schistosomiasis, a parasite carried by snails and transmitted in fresh water. Both infect the gastrointestinal tract and cause diarrhoea and both can cause these wild elevations in eosinophils.

But now Dr Bia hesitated. He was certain that schistosomiasis was found in Tanzania. What about *Strongyloides*? And was there any other bug that could do this? Even though this was his specialty, Dr Bia wanted to be certain that he didn't miss anything. Laboratory cultures of blood and stool could probably provide an accurate identification, but that would take days. And this patient was too sick to wait.

Dr Bia told Dr Hammami he would get back to him. Hanging up, Dr Bia turned to his computer and consulted his own expert – GIDEON. It is an expert system created to help physicians diagnose infectious diseases based on their country of exposure. The program recognises 337 diseases, which are organised by country. Dr Bia opened the Diagnosis module of the program and entered the information he learned from Dr Hammami. He also checked out the Epidemiology module for both strongyloidiasis and schistosomiasis parasites, and then the Therapy module to review the best options for treatment. Within ten minutes he had a plan.

'I used GIDEON to be certain I wasn't missing anything,' he told me later. 'It confirmed my hunch about the best way to proceed.'

Dr Bia called Dr Hammami back. 'Let's just treat her for both parasites,' he said. 'A two-day course of ivermectin for the strongyloidiasis and a double dose of Praziquantel to knock out the schistosomiasis. And before you start the medicine, send blood and a stool sample to our lab.'

Within two days of starting the medications, the vomiting and diarrhoea stopped. The fever disappeared. The patient started to eat. She went home after four days feeling much better, though it would be months before she was completely normal.

The Yale tests showed that the patient had had schistosomiasis. The tiny parasite is carried by a species of East African snail. During heavy rains, snails are washed into rivers, where the parasites disperse. The patient had done some of her research by collecting river water samples. She later admitted that she hadn't worn the protective boots while in the water. They were, she thought, too cumbersome.

Schistosomiasis is such an uncommon disease in the United States that it's not surprising that it was initially missed and the patient misdiagnosed. But the patient might have died before anybody figured it out. Only because Dr Hammami recognised the significance of the abnormally elevated eosinophils, and consulted an expert in infectious diseases, was the correct treatment found. And, in this case, the expert recognised his own limits and consulted a 'digital brain' – an expert system that confirmed his hunches, ruled out other possibilities, and pointed the way to effective therapies.

'I'm not a big high-tech guy,' Dr Bia told me. 'But if you don't know about a particular disease or a particular region, you can miss something. This program helps you narrow down the differential. You can look at diseases in certain countries. If someone has a fever and a rash, and they are just back from Ecuador, you can put in the symptoms and the country and it will come up with a list of possible infections.'

Expert systems such as GIDEON are used at least occasionally today by specialists such as Dr Bia. But most general practitioners don't use such systems – or *any* type of computerised diagnostic decision support. In the case just described, Dr Hammami – a nonspecialist – recognised the clue in the abnormally elevated eosinophils using nothing but his own hard-won medical knowledge. But what about the nurse and doctor who had seen the patient first? This is precisely the kind of situation in which a never-forgetting digital medical brain would seem to be an ideal tool. If the lab results had been typed into a computer program that was 'trained' to watch for anomalies, an alert might have immediately appeared on the screen, prodding the nurse to consider a parasitic infection and reminding the doctor that malaria did not cause a rise in this type of white blood cell.

This, of course, was the vision that had inspired MIT's Peter Szolovits and many others in the 1970s: a computer assistant that was so fast, accurate, and well integrated into the flow of medical information that it would save doctors' time and patients' lives. Such a tool does not yet exist. But with the rise of the Internet, advances in computer speed and memory capacity, and the proliferation of computers throughout the medical system, a second generation of diagnostic decision support systems has been developed that, if not the Holy Grail, has inspired hope that a more perfect system may yet be achieved.

The current paragon of second-generation diagnostic decision support systems was, ironically, the result of a near fatal example of misdiagnosis.

It was early summer, 1999, in suburban London. Three-year-old Isabel Maude had a fairly robust case of chickenpox. Her parents, Jason and Charlotte, brought her to their family doctor even though they weren't at all concerned. After all, chickenpox was an expected childhood rite of passage. The doctor confirmed the diagnosis and sent them home with the standard suggestions for ways to reduce the itching.

But several days after that visit, Isabel developed a high fever, vomiting, diarrhoea, and severe pain and discoloration of the chickenpox rash. Worried now, Jason and Charlotte took Isabel to the emergency room. The doctors examined Isabel and reassured them that her symptoms, while more serious than normal, were not unheard of for chickenpox. They assured the parents that the symptoms would clear within a few days.

The symptoms didn't clear. They got worse. Jason and Charlotte's concern grew into panic. Again they took Isabel to the ER. This time, within a few minutes of her arrival, Isabel's blood pressure dropped dramatically and she required emergency resuscitation. It was suddenly obvious that Isabel was suffering from something a great deal more serious than chickenpox. But what? The doctors had no clue. She was rushed to the paediatric intensive care unit at St. Mary's Hospital in Paddington, London, where Dr Joseph Britto, a paediatric intensive care specialist, took over.

Britto recognised that Isabel was suffering from a rare, but well-described, complication of chickenpox – toxic shock syndrome and necrotising fasciitis – known in the popular press as the flesh-eating disease. To treat the necrotising fasciitis, Isabel underwent an emergency operation to remove the infected skin, leaving extensive scars around her stomach and requiring multiple reconstructive operations. Isabel spent two months in the hospital, including a month in the paediatric intensive care unit. She had kidney failure, liver failure, respiratory failure. Several times her heart stopped and she had to be resuscitated. She hovered on the brink of death for weeks.

Slowly, however, she began to recover. The scars from the surgery are today the only physical reminder of her ordeal. As of this writing, she is a bright and active elementary school student.

For Isabel's father, however, the traumatic events were life-changing. The wrenching emotions of watching his child suffer, and the frustration of seeing her condition misdiagnosed, ignited a passion in Jason Maude to do something to improve the system.

At the time, Maude headed equity research in London for AXA Investment Managers, which oversaw $500 billion in investments. He was familiar with using computers to analyse large amounts of complex data. He talked to Britto about the possibility of using computers to improve medical diagnosis. Britto had already been thinking along the same lines, and in July 1999, the pair formed Isabel Healthcare, with the goal of developing a Web-based diagnostic system for physicians.

Britto was convinced that the risks of misdiagnosis could be solved. He likes to compare medicine's attitude toward mistakes with the airline industry's. It was at the insistence of pilots, Britto frequently remarks – who have the ultimate incentive not to mess up – that airlines have studied their errors and nearly eliminated crashes.

'Doctors,' Britto often adds, 'don't go down with their planes.'

The system that Britto helped develop goes considerably beyond the type of expert system represented by GIDEON. Doctors using the diagnostic tool that Britto and Maude named Isabel can enter information using either key findings (like GIDEON) or whole-text entries, such as clinical descriptions that are cut-and-pasted from another program. Isabel also uses a novel search strategy to identify candidate diagnoses from the clinical findings. The program includes a thesaurus that facilitates recognition of a wide range of terms describing each finding. The program then uses natural language processing and search algorithms to compare these terms to those used in a selected reference library. For internal medicine cases, the library includes six key textbooks and forty-six major journals in general and subspecialty medicine and toxicology. The search domain and results are filtered to take into account the patient's age, sex, geographic location, pregnancy status, and other clinical parameters that are either selected by the clinician or automatically entered if the system is integrated with the clinician's electronic medical record. The system then displays suggested diagnoses, with the order of listing reflecting the degree of matching between the findings selected and the reference materials searched. As in the first-generation systems, more detailed information on each diagnosis can be obtained instantly using links to authoritative texts.

Isabel has had its share of success stories, which the company is understandably proud of. An example occurred not long after Isabel was first available publicly. Dr John Bergsagel, a soft-spoken oncologist at a children's hospital in north Atlanta, read about the new system and asked to be one of the doctors who would serve as beta testers.

On a weekend day not long afterward, a couple from rural Georgia brought their four-year-old son to the hospital's ER. It wasn't their first visit. Their son had been sick for months, with fevers that just would not go away. The doctors on duty ordered blood tests, which revealed that the boy had leukaemia – a type of cancer that attacks cells in the blood. But there were a few things about his condition that didn't add up. For example, the boy had developed these odd light brown spots on his skin around the time these fevers started. No one could figure why these marks appeared but the doctors felt that it wasn't important and scheduled a course of powerful chemotherapy to start on Monday afternoon. Time, after all, is the enemy in leukaemia.

When Bergsagel got the case on Monday, it was just one of a pile of new cases. Reviewing the lab results and notes from the examining doctors, Bergsagel was also puzzled by the brown marks, but agreed that the blood test was clear enough – the boy had leukaemia. But the inconsistencies in the boy's case bothered him. He suspected that, although everyone had made note of the rash, the clear diagnosis of leukaemia may have drowned out any remaining questions.

'Once you start down one of these clinical pathways,' Dr Bergsagel said, 'it's very hard to step off.'

But Bergsagel decided to do just that; he decided to give Isabel a shot. He sat down at a computer in a little white room, behind a nurses' station, and entered the boy's symptoms.

Near the top of Isabel's list was a rare form of leukaemia that Dr Bergsagel had never seen before – one that often causes brown skin spots. 'It was very much a Eureka moment,' he said.

He immediately halted the order to begin massive chemotherapy. The type of leukaemia the boy had was particularly deadly and could not be cured or slowed with any of the chemotherapeutic drugs available. Putting the boy and his family through the pain and rigour of chemotherapy would have been excruciating, potentially deadly, and completely pointless. The only possible cure for this form of leukaemia was another dangerous option: a bone marrow transplant. The procedure was done, even though the chances of a cure were low. The boy lived another year and a half.

Such anecdotes cannot provide proof of the true utility of Isabel. In order to measure how well the program can perform, two researchers (without any financial or other interests in the system) decided to test the system in cases in a more systematic way.

Mark Graber and a colleague tested the system with fifty case studies drawn from the pages of the *New England Journal of Medicine.* Since Isabel accepts information two ways, the researchers tested it in both modes. In one, Graber manually typed in three to six key findings from each case study. On average this took less than a minute. The correct diagnosis was included in the list of possible diagnoses generated by Isabel in forty-eight of the fifty cases (96 per cent). When the text of entire case studies was cut-and-pasted into Isabel (an artificial, but easy, approach) accuracy declined dramatically, with the correct diagnosis appearing in only thirty-seven of the fifty cases (74 per cent).

The authors note that this performance shows that diagnostic decision support systems have evolved significantly since the first-generation systems developed in previous decades. Still, there are many of the same barriers to wide acceptance of the system. Because Isabel and other systems like it are not fully integrated with other medical information systems, data has to be entered into the system by the physician. This is time-consuming and tedious, although Isabel seems to have worked hard to minimise the work involved. Using this system, doctors can describe the patient's symptoms in everyday language. And the machine is smarter, so the amount of detailed information required is much smaller.

But more important, doctors must decide when to use the system. By far the most common diagnostic error in medicine is premature closure – when a physician stops seeking a diagnosis after finding one that explains most or even all the key findings, without asking himself that essential question: what else could this be? If a doctor is satisfied with his diagnosis, he is unlikely to turn to a digital brain at all, and thus the potential value of the system is lost.

So even this new generation of clinical decision-making systems such as Isabel, improved as they are over older programs, is still not widely used. Even Dr Bergsagel, whose use of Isabel so vividly illustrates that system's power, says he uses it only a few times a month.

'The systems available today are still cumbersome to use,' says Jerome Kassirer. 'Doctors still have to input all sorts of stuff into these programs . . . and nobody has the time to type it all in. Besides, most of the time you don't need the system. Most of the day-to-day issues a doctor sees are amenable to the traditional kinds of diagnostic approaches that we've used for years. In fact, it's easier these days because we've got echos and CT scans and MRIs.'

One final impediment exists for Isabel and its competitors: price. Isabel is made available to hospitals on a per-bed cost basis, which works out to about $80,000 for a typical hospital. Individual doctors can buy the service for an annual fee of $750.

Although hardly unaffordable by either institutions or individual doctors, the cost of commercial diagnostic decision support systems means that such programs are vulnerable to competition from what might seem like an unlikely quarter: Google.

## *Googling a Diagnosis*

Patients, friends, and family have periodically confessed to me that they regularly use Google to investigate their own symptoms. My adolescent daughter does it whenever she is baffled by one of her own body's new and peculiar ways. They are not alone in this. According to a 2005 survey done by the Pew Center, 95 million Americans looked for health information on the Internet. I'll bet that most of those people somewhere along the line in their search used Google.

I got an e-mail several years ago from a reader who had managed to diagnose herself using Google when she developed fever and a rash. She didn't start with Google. She started with a man she had always trusted – her doctor.

'I always heard that when your palms itched it meant you were coming into money,' she told her doctor when he entered the exam room. 'No money so far,' she continued, 'but lots of fever.' Dr Davis Sprague eyed her attentively. They'd known each other for years, and despite her playful tone he thought she looked pretty sick.

She'd been well until a few days earlier, she told him. She had a little pain when she went to the bathroom, which made her think she had a urinary tract infection, and so she'd increased her fluids. That didn't work, so the next day she came in and saw a different doctor, who started her on an antibiotic and a painkiller. She didn't get better; in fact, that's when she first noticed the itchy palms. The next morning she was so achy she could barely get out of bed. That night, she had shaking chills and a fever of 102°.

The rash appeared the following day. It started on her arms, her face, and her chest. She stopped taking the painkiller, thinking the rash could be an allergic reaction to it, she told him. But the rash just kept spreading.

Now Sprague was worried. The patient was fifty-seven years old, and other than a back injury a few years ago and some well-controlled high blood pressure, she had always been healthy. Not today. He was glad she was the last patient of the day because he could tell this was going to take some time.

On examination she looked tired, and her face was flushed and sweaty. Her short, dark hair lay plastered to her scalp. She had no fever, but her blood pressure was quite low, and her heart was beating unnaturally fast. The rash that now covered her body was made up of hundreds of small, flat red marks. The newest ones, those on her legs, were like red-coloured freckles. The ones on her arms and chest were larger – the size of nickels – and less well defined. The rash didn't itch or hurt. But the palms of her hands, though rash-free, *were* red and irritated. A urine sample showed no evidence of an infection, but was positive for blood. That might have been a result of the fever, or it could indicate kidney damage.

'You need to go to the emergency room,' Sprague instructed the patient. 'You may even need to be admitted to the hospital. I'm not sure what you've got, but I am pretty sure that this is serious.'

She might have developed an allergy to one of the medicines she was taking, he explained, which could be dangerous and might even require other medications. What he was really worried about, though, was that she had some sort of infection that was spreading throughout her body. In a hospital they could test her blood and get a better sense of what was going on.

The ER doctor ordered what seemed like an endless stream of blood tests as well as a chest X-ray. When all the tests came back normal, he decided she was well enough to go home. It probably was an allergic reaction, he told her, and gave her a different antibiotic. She should follow up with her doctor in a couple of days, he said.

Two days later, she was back at Sprague's office. She did feel a little better, she said, but she was still having fevers, and now she felt short of breath with even minimal effort. 'What do you think is going on?' she asked.

Sprague wasn't sure. Maybe the ER doctor had been right, and it really was an allergy – she was a little better since they'd changed the antibiotics. But the shortness of breath started after that. He was still worried about infection. Fever and rash were common symptoms. It could be a viral illness – Coxsackie? West Nile? Or was it bacterial? These symptoms, he told her, were so nonspecific that they could be found in everything from garden-variety Lyme disease to something really exotic like Rocky Mountain spotted fever. 'We may never figure it out,' he confessed. But since she was getting better, he was willing to give her a few more days. If she was still spiking fevers then, he'd send off some blood work to try to find an answer.

At home, though, the patient continued to worry. That night she sat down at the computer to do a little research of her own. 'Rash, adult, fever,' she Googled.

When you Google a set of symptoms, you don't necessarily get the most common or the most likely diseases; you get the diseases with the greatest number of links from other Web sites. Her Google search brought up dozens of fairly unusual, but well-linked, illnesses: coccidioidomycosis – a fungal infection most common on the West Coast; dengue fever – endemic to the tropics and near tropics; measles; scarlet fever.

But the patient immediately focused on the first result: Rocky Mountain spotted fever, which her doctor had mentioned. As she read about the disease, she began to feel a little panicky. The description of the symptoms, she said, fit her perfectly: the rash, the fever, the muscle aches. The rash, she read, can involve the palms of the hands, which is pretty unusual. She didn't have a rash there, but her palms were red and itchy. Also, the disease is transmitted by dog ticks – she had a dog. It's most common in the summer – it was August. Though it's rare, it is more commonly seen on the East Coast than in the Rockies, and she was in upstate New York. People can die from this disease, she read. It's the deadliest of all the tick-borne illnesses.

She called the emergency room where she had been seen. Had they tested her for Rocky Mountain spotted fever? No, she was told, why would they? They had never seen a single case in the area. She hung up feeling somewhat relieved. They didn't think it was Rocky Mountain spotted fever; Dr Sprague didn't think it was. Chances are that it wasn't.

Over the next few days, the patient started to feel almost normal again. The rash was fading – though now it itched like crazy – and her energy was coming back. But she continued to have fevers at night and still occasionally felt short of breath. She returned to Sprague's office one more time. 'I'm glad to hear you're feeling better, but these fevers worry me,' he said. 'I want to send off some tests.'

'What about Rocky Mountain spotted fever?' the patient asked. She confessed that she had looked it up on the Internet and thought the symptoms were close to what she had. The doctor thought for a moment. 'I don't think that's what you have, but let's add it.' He had heard doctors complain about their patients surfing the Web for diagnoses, but he didn't mind. He had never seen Rocky Mountain spotted fever – maybe she was right.

The results came back a few days later. 'You're an internist's dream,' the doctor said with a smile as he entered the exam room. 'It really *is* Rocky Mountain spotted fever, and I would have completely missed it if I hadn't listened to you.' He started the patient on doxycycline – the antibiotic of choice for this bacterium. Her body seemed to be fighting off the illness without it, but he wasn't taking any chances. Within a few days her fever was gone, the rash was fading, and her palms were beginning to feel normal.

I asked the patient how she felt about her doctor, who had come so close to missing this diagnosis. 'But he didn't miss it. He was the first to think of it. And he sent off the test – even though it could prove him wrong. He just wanted to figure out what was going on.'

This case illustrates a real and growing trend – patients who either diagnose themselves by using the Internet or follow up on their doctor's diagnosis in that manner. But it's not just patients using the power of Google and other search engines these days. A doctor wrote to the *New England Journal of Medicine* about an amazing diagnosis made at his institution. The case involved an infant with diarrhoea, an unusual rash, and multiple immunological abnormalities. The patient was discussed at length in a case conference with residents, attending physicians, and a visiting professor. No consensus was reached. The letter continues:

> Finally, the visiting professor asked the fellow if she had made a diagnosis, and she reported that she had indeed and mentioned

a rare syndrome known as IPEX (immunodeficiency, polyendocrinopathy, enteropathy, X-linked). It appeared to fit the case, and everyone seemed satisfied . . .

'How did you make that diagnosis?' asked the professor. Came the reply, 'Well, I had the skin biopsy report, and I had a chart of the immunologic tests. So I entered the salient features into Google, and it popped right up.'

This story and their own experiences with patients who had consulted the Internet for information about their own symptoms prompted a pair of Australian researchers to test Google's diagnostic accuracy.

Like Graber, they used the medical case studies published in the *New England Journal of Medicine*, selecting three to five keywords from each article, and entered them into Google before they, themselves, read the actual diagnosis. The doctors selected and recorded the three most prominent diagnoses that Google came up with for each case. Then they compared the Google findings with the real diagnosis.

The result? Google flunked. Google found the right diagnosis for only fifteen out of twenty-six cases (58 per cent). Of course, Google isn't designed to provide diagnostic support for doctors, so any right answers provided by the powerful search engine are bonuses. One interesting observation was made by the authors: Google was most accurate for diseases that had unique signs and symptoms or rare presentations. This isn't surprising to any of us who use Google, but it's interesting. As anybody who has used a search engine knows, the more unusual your target is, the easier it is to find. For example, if you want to Google two friends, you are much more likely to find the one named Ionia Khammouane than the one named Ann Jones. Information on Ionia is going to pop right up, just like the diagnosis of the case of the child with leukaemia and the brown marked rash.

What's interesting is that it's precisely the unusual disorders – the ones with peculiar symptoms that doctors rarely see – that can be most baffling to both doctors and patients. In the case I presented in an earlier chapter, a resident in our program was able to diagnose a patient with intermittent nausea and vomiting because of an unusual symptom – her nausea was improved by hot showers. By Googling that, Amy Hsia was able to identify an unusual and recently described disease called cannabinoid hyperemesis.

Because Google is so universally available, simple, fast, and free, it may become the go-to diagnostic aid for oddball cases. Even the august *New England Journal of Medicine* finds Google 'helpful in diagnosing difficult and rare cases.' Google gives users ready access to more than three billion articles on the Web and is far more frequently used than PubMed for retrieving medical articles.

The authors of the Google study note that, in fact, Google is likely to be a more precise diagnostic tool for clinicians than the lay public because clinicians will use more specific search terms ('myocardial infarction' rather than 'heart attack,' for example) and will be better able to identify likely hits because of their preexisting knowledge. Patients, using everyday language, are likely to end up with fewer useful hits buried in pages of irrelevant sites. Their ability to distinguish the useful hits will be compromised by their unfamiliarity with medical language.

The power of Google in the realm of medical diagnosis has not been lost on Google itself. Google has formed a Health Advisory Panel to inform its work in this area. And Google has launched a major effort to improve the quality of medical-related searches by having reputable organisations (such as the National Library of Medicine) and individual doctors flag Internet sites offering reliable information. These sites are then given prominence when search results are returned and are labelled with the individual or organisation that has vetted them.

Google is very open about its plans to improve search capabilities for patients, but the company is mum on the subject of doing the same thing for physicians (Google representatives declined to be interviewed on this subject). Perhaps that's because doctors are a valuable audience and if Google can find a way to improve diagnostic search results to the point of being more accurate than Isabel and other commercial systems, it could effectively capture the market and be able to leverage all those physician 'eyeballs' with advertisers.

But even a more accurate Google-based diagnostic decision support system wouldn't really solve the problem of missed diagnoses. To begin with, any system that must be consulted separately from the digital workspace in which a doctor or nurse deals with a patient will only be used when there is uncertainty in the mind of the health care professional. If a doctor is sure of her diagnosis, or a nurse is certain that the correct medication has been prescribed, they won't turn to Google (or Isabel, or DXplain, or any other system).

Computer programs won't really make a dent in the problem of misdiagnoses and other types of medical errors until they are much 'smarter' and easier to use than they are today.

'Future systems need to operate in the background,' says Eta Berner, the researcher who has tracked progress in medical computing for decades. 'The doctor shouldn't have to enter anything. The system should be able to extract information from what the doctor or nurse is already doing . . . taking notes or entering lab values or prescribing medications. The system should be intelligent enough to provide an alert or a reminder only if something is really missing . . . a test, for example, or a medication.'

Berner foresees a time when all of the now fragmented information streams in the health system will be unified and made consistent. Patients' health records will be fully digital – including images such as MRI scans or X-rays. Standard words, phrases, and units of measurement or description will be used so that computer systems in distant locations can intelligently and accurately use the information. Doctors and nurses will enter all information in digital form – handwriting (never doctors' strong suit anyway) will be obsolete.

With this kind of a system in place, the possibility of infection with the schistosomiasis parasite would have popped up the very first time the young woman described earlier was evaluated in an emergency room. The likelihood that little Isabel Maude was suffering from a rare complication of chickenpox would not have been easy to ignore. And the patient with Rocky Mountain spotted fever wouldn't have had to use Google herself . . . her doctor would have already seen the tight fit between her symptoms and that possible diagnosis.

Of course it will be years – and more likely decades – before this kind of a system is in place. And although I think it is inevitable that the vast resources of the digital age will become more fully integrated into our health care system and the doctor's diagnostic routine, it may not take the form we anticipate. Computers have already revolutionised our diagnostic abilities dramatically. I think the first and most important digital diagnostic tool developed was the CT scanner. It was the development of powerful computers that allowed us to capture data from a series of two-dimensional images to create a three-dimensional representation of the body. Since 1972, when the CT scan was first developed, this tool has made routine diagnoses that would previously only have been discovered after death. So while we envision a future where the computer learns how to think like a doctor, it is possible that its greatest contributions will take a very different form.

Would a kind of super-efficient, integrated, intelligent computer system eliminate all diagnostic challenges? Would it replace doctors? Hardly. I believe the process of diagnosis will be made more effective and that it will be faster and easier in the future to zero in on what's really wrong with a patient. But there will always be choices to make – between possible diagnoses, between tests to order, and between treatment options. Only a skilled and knowledgeable human can make those kinds of decisions.

And, of course, people need more than the right treatment for the right disorder. They need to be heard, they need reassurance, explanations, encouragement, sympathy – the full range of emotional support that is a critical part of what we doctors try to do: heal.

# AFTERWORD

## *The Final Diagnosis*

'I'm sorry,' the young man on the telephone said to me. His voice was hushed and sympathetic, difficult to hear over the usual commotion of the clinic bustling just outside my office door. He was a stranger to me. He said his name was Jorge. He was an old friend of a young woman we both knew quite well. 'I'd chatted with her on the phone maybe twenty minutes earlier. She said come by and so I just drove on over.'

He told me that he'd rung her bell early that sunny September morning and when there was no answer, he clanged through the backyard gate. When he saw her stretched out on the chaise longue in her bathing suit, his first thought was how pretty she looked. 'I'm a married man, so it wasn't like that, but she's always been a looker.' When she didn't reply to his 'Hey, how's it going?' he approached her and put his hand on her shoulder. Her skin felt warm but he noticed how strangely pale she was under her tan. 'And I knew then, I knew. Her cell phone was right there next to her, like it always was, so I picked it up and dialled 911.'

I thought back to the last time I'd seen Julie: her tanned cheeks still unlined, her eyes so blue that even the whites were the colour of robin's eggs. I could hear her deep tobacco-coarsened drawl and her earthy sense of humour. I closed my office door and dropped into my chair.

My beautiful and mysterious little sister was dead.

My first thought, when thought was finally possible, was how? More than anything, I wanted to know how a young woman could die so suddenly that she didn't even have time to call for help. What happened?

It was a strangely familiar question. When patients of mine have died, their spouse or parent or child or friend would ask me this very question after I broke the news. In waiting rooms outside the emergency room or ICU, shocked, sad, crying – they would ask: Doctor, how did this happen? How did this person, so very alive not so long ago, die? I would do my best to answer, to pull together the strands of a devastating illness or collapse, but it seemed a peculiar question – as if an explanation could somehow soothe the jagged edges of loss. But it made sense to me now. I suddenly understood that terrible need to know how.

At forty-two, my sister was healthy. But she was also an alcoholic. For the past fifteen years or so, her life had been dominated by this desire, and then this need, to drink. She'd started out – like so many – with excesses in high school, but calmed down after marriage and the birth of the son she loved. Over time, and for reasons I will never know, Julie's drinking became more frequent. Weekend binges rapidly became the daily dose she'd sneak as she got her son ready for day care, or as she set out for work, as she prepared dinner or put her son to bed.

She tried to stop. Again and again she would check herself into a hospital, or simply start going to AA meetings and try – I think, really try – to stop. She would call us almost daily, triumphant with the exact number of days, even hours, since her last drink. Then the calls would become less frequent. Her voice mail would tell us she'd call us back but she rarely did. And then finally there would be silence. Until she would try once more. My sisters and I – we were a family of five sisters – watched in helpless distress. Over the years we'd learned what all relatives of alcoholics learn: that everything we could do still was not enough.

And then she died, as mysteriously as she had lived.

What could kill a young woman that young, that fast? Jorge had found her cell phone along with a pack of cigarettes and a Coke sitting right beside her. She was obviously tanning herself, relaxing in the summer sun. Whatever killed her struck so quickly that she could not reach over to pick up the phone and dial 911. What could do *that*? I couldn't get that terrible question out of my mind. As I made arrangements to travel home, I puzzled over it. I went into my doctor mode – in part because it was a way of managing my grief and in part because it's what I'm trained to do. And without really wanting to, I found myself putting together a differential diagnosis, searching for scenarios that might explain how my sister had died so abruptly.

Certainly a heart attack can be quick and deadly, especially at a young age. But that would be unusual in a forty-two-year-old woman. And we had no family history of heart disease. A ruptured blood vessel in her brain could cause an instantaneous loss of consciousness and rapid death. A massive clot that went to her lungs was another possibility. She was a smoker; maybe she was also taking birth control pills. That combination has been linked to such clots. Infection seemed unlikely. And yet, had she been sick? I didn't know. Suicide was unthinkable to me, but it had to remain a possibility. She was often deeply depressed during these relapses. An accidental overdose was also possible.

The coroner in Savannah, Georgia – where she had lived her last year and where she had died – ordered an autopsy be performed. Although one of my sisters was upset with what she saw as a violation, I was grateful. An autopsy, I hoped, would provide me with this necessary and final diagnosis.

Autopsy – the word comes from the Greek *autopsia*, meaning to see for oneself. Historically the autopsy has played a critical role in medicine. For centuries everything we knew about disease was derived from examining the body after death. Even now when my patients ask me about their aches and pains for which I have no diagnosis, I confess to them that our knowledge of diseases that *can't* kill you is fairly new and much less developed because even now most of what we know about disease was derived postmortem.

Medicine's first toehold into modern-day diagnosis came at the last half of the eighteenth century, when Giovanni Battista Morgagni, a physician and professor at the University of Padua, published *On the Seats and Causes of Diseases Investigated by Anatomy*. This book, completed when Morgagni was seventy-nine years old, was composed of hundreds of beautifully detailed drawings from autopsies that he'd performed over the course of a long career. These carefully drawn images revealed the destruction and distortions of the anatomy hidden beneath the skin and leading to death. By showing exactly how disease manifests itself in these visible, concrete ways within the body, the work inspired generations of doctors to investigate the process by which disease can distort and derange our most fundamental anatomy. For centuries disease and death had been attributed to humours or spirits or other intangibles and not something as real, or as clearly visible, as it was in these images.

For the past 250 years autopsy has been one of medicine's most reliable sources of information about the nature of disease. Cancer, heart disease, haemorrhage were all first seen through the exploration of the body after death. In the twentieth century, autopsy was used as the ultimate diagnostic tool. At its peak, up to half of all patients who died in the hospital underwent postmortem evaluation. Too late to help the patient, what was revealed was often useful knowledge for the doctor, the hospital, the family. Diseases missed or undetectable with the available technology were finally made visible. Doctors could use the knowledge for the benefit of their next patients. Hospitals used the information as a form of quality assurance on the care they provided and the skills of the doctors who practised there. There were benefits for the bereaved family as well. The disease that took their loved one could be a risk for them as well.

These days, patients who die in a hospital rarely make it to the pathologist's table. Hospitals used to be required to perform autopsies. The Joint Commission on Accreditation of Healthcare Organizations – the regulatory body overseeing hospitals – required these institutions to maintain autopsy rates of at least 20 per cent (25 per cent for teaching hospitals), which was, and continues to be, the rate most advocates say is the minimum for monitoring diagnostic and hospital error. The commission eliminated that requirement in 1970. Medicare stopped paying for those that still got done a few years later.

Until quite recently autopsies were also considered an essential component in medical training. Residency programmes were required to get autopsies on 15 per cent of all the patients who died while under resident care. Seeing the real ravages of disease was considered an important part of medical training. But the requirements for most medical trainees were rolled back in the 1990s. Small residency programmes objected to the ever growing cost – autopsies were not paid for – and enforcement of the rule was difficult.

Even before the rollbacks of the requirements on hospital and training programmes, the number of autopsies performed had plummeted. In the 1960s, nearly half of those who died in the hospital were autopsied. Only forty years later, at the turn of the twenty-first century, that rate had dropped to less than six per one hundred in-hospital deaths. We don't even know how many are done now because that data isn't collected anymore. In the community hospital where I take care of patients, there were ninety-three autopsies done in 1983. One recent year, we had performed a grand total of eleven autopsies and almost half of those were on stillborn infants.

What's happened here in the United States has happened everywhere. There's been a global decline in the rate of autopsies – a reflection, in part, of the increased cost of health care, augmented by long-standing cultural concerns about this kind of violation of the body. But the real driving force behind this plunge has been the growing confidence of doctors and patients that the diagnoses given in life were accurate.

Certainly a doctor's ability to make an accurate diagnosis has improved dramatically over the past half century. A recent study done by the US Agency for Healthcare Research and Quality suggested that the likelihood that a doctor will make an important diagnostic error has declined by 25 per cent each decade since the middle of the century. It is a testimony to the effectiveness of the new technology of testing we have at our fingertips.

But that study also shows that doctors still miss important problems. Of the few autopsies still done, a diagnosis that could have changed the management of the patient – and therefore possibly changed the final outcome – was found in one out of twelve autopsies. These days, doctors only order autopsies when the patient's death came as a surprise or the underlying illness was not understood. Given that, it's perhaps not surprising that something important was missed; it's why the doctor got the autopsy in the first place. And yet several studies have shown that doctors are unable to predict which cases will provide the surprises. It turns out that in medicine (as in war, according to Donald Rumsfeld) there are the things you know you don't know, and then there are the things you don't know that you don't know. Autopsies are one way to explore those dark recesses. The drop in the number of autopsies suggests that neither doctors nor hospitals are interested in exploring the deep recesses of what we don't know we don't know.

My sister didn't die in the hospital, where the odds that she would ever have a final diagnosis were small. She died 'in the field' and so hers became a medicolegal death. The medical examiner and coroner are twin investigative arms, designed to look into unexpected deaths. The most important difference between the systems is that medical examiners are always physicians, usually pathologists, appointed by the state; a coroner is an elected officer, and rarely a physician. Both are charged with the investigation of any unexpected death outside the hospital. As watchers of *CSI* know, detecting whether a crime occurred causing the death is the primary goal. In addition, medical examiners can provide a public health service – an early alert system to identify emerging infections. Because my sister died in her own backyard, she fell under the authority of the state of Georgia's coroner's system and so her body was taken for autopsy. The unexpected death of a young woman merited an investigation – one that I hoped would provide me with an answer.

As we waited for the coroner to finish his gruesome investigations, I continued to try to find out more about the hours and days before she died. Were there clues there? Jorge, the friend who'd found her, provided a few details. They were painful to hear. My sister had been on a binge over that

Labor Day weekend. A serious binge. She'd called him that morning, filled with remorse and shame but also determined that this time she would be able to stop. She felt weak, tired, achy. She had a stomach ache, a headache; her back hurt. He said he'd be right over, and he had been. And that's when he found her.

Another sister had spoken to her just a couple of days before she died. 'She went to the doctor last week, and she never does that. She had a stomach ache. But the doctor didn't find anything. Anyway, I wonder how much she even told him.'

I called the office where she'd been seen. 'She was here once, several years ago, and then again about a month ago,' the doctor reported. I could hear papers rustling as he paged through her chart. 'During that visit she complained of some persistent lower abdominal pain for the past few days. Some nausea, some vomiting, no diarrhoea. She denied any past medical history, took no medications. On physical exam she was a thin, tired-appearing woman. Her blood pressure was normal 122/80, heart rate was high but still in the normal range. She had no fever. Her abdominal exam was unremarkable: minimal generalised tenderness, bowel sounds were present. I didn't do a rectal.' Pages crinkled. 'A urinalysis was normal. A CBC [complete blood count – a test that quantifies white blood cells, red blood cells, and platelets] showed no evidence of infection. I thought she might have had a virus and I gave her something for her nausea and a mild painkiller. I told her to call if she didn't get better.' He paused and the rustling stopped. 'I didn't know she died. I'm sorry.'

I flew home to our family graveyard, already crowded with the stones of the last generations. My sisters and I received flowers, condolences, casseroles. We waited for the coroner to send us her body and when it was delivered, we buried her. People came from our hometown and her new town. I met Jorge and a few of her other friends from AA. I found then that we all struggled with the same question: how?

After the funeral I called the coroner's office, confident that they would have an answer. The report wasn't complete – laboratory data was still pending – but I persuaded the office assistant to stumble through the report to the conclusion. They had completed the autopsy but had found nothing, no evidence at all of what had killed my sister. The woman on the phone was kind, and apologetic. She could feel my disappointment.

I went to my first autopsy as a first-year medical student. I had half a year of anatomy under my belt, so I had seen death up close before. There was a small group of medical students and residents there to observe. As we put on the paper jumpsuit, face shield, and mask that are required in an autopsy room, the pathologist briefly outlined the case. It was a young woman who had died just days after giving birth to her first child. The last weeks of her pregnancy had been complicated by high blood pressure – too high to control even with the several medicines that she had been given. She then developed kidney and liver failure and was diagnosed with preeclampsia – a mysterious and unusual complication of pregnancy. The only successful treatment for this is delivery of the baby, and this young woman had had a caesarean.

But even after this child was delivered the mother remained ill, and then suddenly died. What had killed her? That was the question the autopsy was to answer.

We trooped into the autopsy room, a large, brightly lit chamber with institutional green walls and dotted with several body-length stainless steel tables. At each station there was a scale, a table for specimens, and a hose trickling water along a trough beneath the table. The deep rumble of an exhaust fan added to the industrial feel of the place.

Despite the thick paper mask I had fastened over my nose and mouth, the sickly sweetness of the cleansers and preservatives was apparent, and beneath that the fetid animal smell of blood and stool. The body of the young woman lay on the table. She was naked – tiny and vulnerable on this long cool slab. She could have almost been asleep except for the mannequin pallor of her skin. Her short brown hair hung down to the table; her neck was elevated on a block of wood. A small tattoo on her shoulder showed a bird in flight.

The technician announced the time and then, with practised swiftness, picked up a scalpel and inserted the blade into the young woman's chest just beneath the left collarbone. He sliced down and across the chest to the bottom of the middle of the rib cage. No blood flowed from this wound.

He swiftly cut through the ribs on the right, completing a large V across her chest, then continued straight down her abdomen, past the still raw surgical scar from her C-section down to her pubic bone. The calm, utilitarian brutality was fascinating and a little repulsive. Still, the laboratory-like environment and the subtle changes in the body that screamed that no life was left in this shell made the unthinkable possible.

The technician, a middle-aged man with beefy arms, opened the chest and abdomen, revealing the organs within. One by one the organs were cut free of their connections, brought out of the body, inspected, and then weighed. Every observation and measurement was announced and recorded, to be transcribed later.

The lungs were lifted out to reveal the heart, which, we were told, was enlarged. She was so small that it looked tiny to me but when it was weighed, there was a murmur among the cognoscenti, an acknowledgement that the heart was indeed surprisingly large. The rest of the organs were removed, inspected, and weighed, then lined up on the table for closer inspection later.

The technician moved up to the head. He made an incision across the back of the scalp, then peeled the tissue forward as easily as you might fold back the skin from a banana. Using what looked like a power saw, he quickly cut a circle in the top of the skull. He pried the loosened lid of skull bone away with a slender crowbarlike tool. The pale greyish tan ripples of the brain I knew from my own explorations in anatomy class were not there. Instead I saw what looked like a smooth grey ball, blotched with coaster-sized circles of shiny brown-black. The brain was hugely swollen. The coasters were old blood congealed on the surface. Clearly some large blood vessel in her brain had ruptured, filling all the available space and squeezing the brain to a shiny unnatural smoothness. She'd had a cerebral haemorrhage – a consequence of the high blood pressures that even the birth of her child and all of our medicine were unable to bring down.

When the coroner's assistant told me that my sister's autopsy was unrevealing, I thought about that young woman. Involuntarily I pictured my sister lying on that aluminum slab, the deep blue eyes closed, the sun-bleached hair matted around her, her innermost recesses exposed to the expert eye of those who didn't even know her. It hurt to imagine it. Surely they'd seen traces of the hard life she'd led: dark lines in her lungs revealing her long history of tobacco; an enlarged liver – or perhaps a liver scarred and shrunken from her years of drinking. There was a painful kind of embarrassment as these technicians learned the secrets of my little sister's life. As if they'd walked in on my sisters and me in grief and had somehow seen all our secrets as well. Yet nothing they learned would account for her sudden and unexpected death. I hung up the phone and took a few deep breaths.

These disappointing results actually did have something to tell me. The autopsy would have shown if she'd had a massive bleed somewhere. Or a large clot in her heart or her lungs. Or a deadly infection. Instead, she appeared to be completely normal.

There are only a few things that can kill you without leaving a mark. Had she overdosed on drugs? Alcohol was her drug of choice – did she add anything else to the mix? And if she had, did she do it on purpose? The thought of a despair leading her to take an intentional overdose was almost more than I could bear. The police hadn't found any pill bottles or evidence of illegal drugs at the scene and there was no note. Or could she have had an abnormal heart rhythm? And if she had, what could have caused it? The next step would be for the coroner to examine her blood and tissues for causes that would be invisible to the eye.

The last time I spoke with my sister was on her birthday. I could tell she'd been drinking because she didn't want to talk. 'What's new?' 'Nothing much,' she reported. 'Same old, same old. Going to work, going to meetings, going home.' She took a deep drag off her cigarette. 'How about you?' she asked, avoiding any real talk about her life. I told her a bit about my two kids and we ended our brief conversation, with dissatisfaction at both ends. She said she was going to meetings, but if she hadn't been drinking she would have been full of details, of stories, of humour. My sister was a cheerless drunk: secretive, defensive, quiet; so different from the exuberant, down-to-earth woman she'd been before drinking had taken over her life.

As we cleaned up after the funeral reception, my sisters and I talked about her last few years. The sister who had remained closest, both geographically and emotionally, recalled taking her to the hospital once before. 'You remember, don't you? She was vomiting up blood and I took her to Roper. They took some of her blood and after she was 'scoped, a young doctor came in to see her. He told her that her potassium was dangerously low and they had to give her potassium in her veins.'

Low potassium – hypokalaemia – is a well-described complication of alcoholism. When taken in excess, alcohol can cause the body to dump certain electrolytes – like potassium, like magnesium. Normally this would not cause a problem because we replace these electrolytes every day. Most of us eat far more than our bodies can ever use. But alcoholics sometimes don't replace these vital chemicals. And once these key electrolytes get outside the normal range, it's hard for our bodies to work well. If they get too far from normal, then they can't work at all: our heart simply stops and we die.

Our bodies are well protected against this, normally. But for my sister these were not normal times. Could this critical imbalance have occurred again? The circumstances were right: she'd been on a binge, and probably hadn't been eating. I knew that in the past she'd lost five, even ten pounds while on a binge because she simply didn't eat. I'd forgotten about her history of hypokalaemia. That had happened right after a binge too. Without potassium your heart could just stop beating. No pain; no time to reach for the phone. Could that be what killed her?

After several weeks the coroner was finally able to release her report. No abnormalities were found other than those normally seen after death. There was alcohol but no poison, no drugs, no sign of infection. Her electrolytes were completely out of whack. Her potassium was not too low – as I had expected – but much too high. I called the pathologist who had done the autopsy. Could my sister have died from this unanticipated elevation in her potassium? No. She told me that the high potassium I saw was due to the changes that occur in all bodies after death. If there had been a critically low level of potassium or some other vital chemical, which ultimately made her heart stop, death itself had erased all the evidence.

So the autopsy didn't have the answer. And yet, putting it all together – her history of hypokalaemia, her unrevealing autopsy, the suddenness of her death, I knew what had happened. I could put the story together in my head. Jorge told me that Julie had been drinking and I knew that she didn't eat when she was on a binge. That combination would account for the abdominal pain that sent her to the doctor's office. Her potassium was low. That's why she'd felt so achy and tired the morning she died. The low potassium must have tripped her heart into a fatally irregular rhythm. Her death would have been almost instantaneous – leaving no time to even call 911.

I spent the following Christmas with my three sisters. In a rented beach house on a cold grey December night, after children and husbands had gone to bed, we sat and talked about Julie. Although more than a year had passed, the loss was still fresh, and this holiday – our first together without her – made the pain even sharper. For them, the peculiar facts of how she died were just more of the jumble of unconnected mysteries that so often trailed my little sister. So I told them in plain words about what my textbooks call hypokalaemia, and explained my version of the story of Julie's death. With this final piece of the puzzle in place, it became easier to fit the story of her sudden death into the longer story that we already knew – the story of Julie's disease, the story of her alcoholism, and then into the even longer story of her life. Yes, she was the drinker who died, but she was also the funny, earthy woman whose biting sense of humour helped her handle the toughest breaks tossed her way with a wink and a wicked one-liner.

'You know Julie would just laugh if she could see us now,' one sister remarked dryly, dotting her tears on a ragged tissue. 'She always said that it's not really Christmas until everybody cries. We stay up too late, eat too much, drink too much, see too many people we love and hate. Just too much going on for the human heart to handle.' And then, suddenly, we were able to start trading our accounts of that Julie. She had a way of laughing about the mundane suffering of everyday life that I envied. It felt good to miss her, that much, with all my sisters, and in this way.

We kept up the laughter and the stories until the approaching dawn signalled that it was time to wrap it up. By then medicine wasn't a solace, or even part of the evening. That version of the story had long ago drifted into the deep background of what we all knew now. The chilly, precise language of potassium and arrhythmia had been aired out, unpacked, and retranslated back into the comfortable idioms families speak when the medical personnel have long since left the room. Ultimately, medicine can't bring comfort, but it does help tell the final story in a life. Knowing how someone died makes it easier to remember how they lived. And after medicine has finished doing all that it can, it is stories that we want and, finally, all that we have.

# ACKNOWLEDGEMENTS

This book originated in the pages of the *New York Times* magazine and was only possible because Paul Tough, an editor there, believed that the stories I told in casual conversation could be successfully translated onto the pages of the magazine. Thank you, Paul, for your vision. Over my years there, I have been the beneficiary of the generous guidance of many great editors. Thank you Dan Zalewski, Joel Lovell, Catherine Saint Louis, Ilena Silverman, Katherine Bouton, and Gerry Marzorati.

To the patients who shared with me some of the most terrifying moments of their lives – those hours, days, sometimes weeks between the time when mysterious symptoms appeared and the correct diagnosis finally made – I owe an incalculable debt of gratitude. I have learned so much from you all. Thanks also to the doctors who allowed me to see and recount the uncertainty they faced as they tried to unravel the mysteries of these patients. The diagnostic process is much more than the triumphant declaration of the cause of an illness, and I am deeply indebted to the doctors who allowed me to map the landscape of that uncertainty.

With all these wonderful stories at hand, I was shocked by the challenge of shaping them into the book I wanted to write. Mindy Werner nursed this inchoate mass of ideas and stories into the foundation of this book. Steve Braun used his considerable skills as a reporter to help me find just the right building materials. And Karl Weber, thank you for helping me shape these chapters into the book it is now. My running partners Elizabeth Dillon and Serene Jones listened as I struggled through these chapters as we took on the hills of East Rock. No matter how breathless, they could always be counted on to ask the questions that needed to be asked. Anna Reisman, Eunice Reisman, John Dillon, Pang Mei Chang, Betsy Branch, and Allyx Schiavone read through these chapters more times than I can count – and without complaint. Their comments steered me back whenever I wandered deep into medical arcania, and my stories are better told because of their help. At Yale, Steve Huot, Julie Rosenbaum, August Fortin, Donna Windish, Andre Sofair, David Podell, Michael Green, Dan Tobin, Steve Holt, Michael Harma, Jeanette Tetrault, Jock Lawrason, and the rest of the faculty, staff, and residents created a stimulating and supportive community in which to do this work. Tom Duffy, Frank Bia, Nancy Angoff, Asghar Rastegar, Patrick O'Connor, Majid Sadigh, and Eric Holmboe taught me almost everything I know about being a doctor and helped me shape many of the ideas in this book. The resident reports presided over by Jerome Kassirer were models of clear medical thought and great storytelling. I paged through my notes of these hours of medical exegesis frequently as I was working through these chapters – especially those on thinking.

Jake Brubaker, Edmund Burke, Laura Cooney, Onyi Offor, Valerie Flores, Marjory Guerra, Jason Brown, and Clayton Haldeman, provided an enthusiastic cheering section each week as I slowly made my way through the writing of this book. Paul Attanasio had a vision for how stories like mine could be told on television. Thank you for inviting me into the miraculous world of television doctoring. Thanks also to David Shore – who tapped his inner House to bring life to the doctor-detective Gregory House and his passionate pursuit of diagnosis, which made this topic so near and dear to my heart part of the national conversation.

Charles Conrad, my editor and guiding light at Broadway Books, believed in this book from the beginning. His quiet wit, vision, and (thank God) patience, provided the kind of steady hand I needed throughout. Copy editor Frederick Chase had an eye for detail that prevented any number of embarrassing errors. My friend and agent Gail Ross was certain this was a book well before I was, and held my hands through it all. Gail, I owe you big. Thanks also to Jennifer Manguera, who worked hard to keep my literary house in order.

Finally, I am grateful to my daughters, Tarpley and Yancey. You have been the centre of my world and the gravity in my solar system. When the orbit of this book took me to the darkest part of my own personal universe, your love pulled me back to the warmth of this wonderful family I managed to be part of. And to Jack, without whom none of this would have been possible – which is why this book is dedicated to you.

# *NOTES*

***Introduction: Every Patient's Nightmare***

xv **'cookbook medicine':** Berner E, Graber M. Overconfidence as a cause of diagnostic error in medicine. *Am J Med.* 2008;121:S2–23.

xxii **'an inferential process, carried out under conditions of uncertainty':** Kassirer J. Teaching problem-solving: how are we doing? *N Engl J Med.* 1995;332:1507–1509.

xxii **Institute of Medicine released a report on the topic:** Kohn LT, et al., eds. *To err is human: building a safer health system.* Committee on Quality of Health Care in America, Institute of Medicine, National Academy Press, Washington, D.C., 2000. Book text is available online at http://books.nap.edu/openbook.php?isbn=0309068371.

xxii **Depending on which study you believe:** Graber M, et al. Reducing diagnostic errors in medicine: what's the goal. *Acad Med.* 2002;77:981–999. Holohan TV, et al. Analysis of diagnostic error in paid malpractice claims with substandard care in a large healthcare system. *South Med J.* 2005;98(11):1083–1087.

xxii **Studies suggest that between 10 and 15 per cent:** Berner E, Graber M. Overconfidence as a cause of diagnostic error in medicine. *Am J Med.* 2008;121:S2–23.

xxii **In a study of over thirty thousand patient records:** Leape L, et al.

The nature of adverse events in hospitalized patients: results of the Harvard Medical Practice Study II. *N Engl J Med.* 1991;324:377–384.
xxii **And while postmortem studies:** Goldman L, et al. The value of the autopsy in three different eras. *N Engl J Med.* 1983;308:1000–1005.
xxiii **A study done... in Switzerland:** Sonderegger-Iseli K, et al. Diagnostic errors in 3 medical eras: a necropsy study. *Lancet.* 2000;355:2027–2031.
xxiii **Another study done for the Agency:** Shojania K, et al. The autopsy as an outcome and performance measure. Evidence Report/Technology Assessment no. 58 (Prepared by the University of California at San Francisco–Stanford Evidence-Based Practice Center under Contract No. 290-97-0013), AHRQ Publication no. 03-E002. Rockville, MD, Agency for Healthcare Research and Quality, October 2002.

***Chapter 1: The Facts, and What Lies Beyond***

6 **Indeed, the great majority of medical diagnoses:** Hasnajn M, Bordage G, et al. History taking behaviors associated with diagnostic competence of clerks: an exploratory study. *Acad Med.* 2001;76:10:S14–S16. Hampton JR, et al. Relative contributions of history taking, physical examination and laboratory investigation to diagnosis and management of medical outpatients. *BMJ.* 1975;2:486–489.

6 **In recordings of doctor-patient encounters:** Beckman HB, Frankel RM. The effect of physician behavior on collection of data. *Ann Intern Med.* 1984;101:692–696.

6 **In one study doctors listened:** Dyche L, Swiderski D. The effect of physician solicitation approaches on ability to identify patient concerns. *J Gen Int Med.* 2005;20:267–270. Marvel MK, et al. Soliciting the patient's agenda: have we improved? *JAMA.* 1999;281:283–287. Rhoades DR, et al. Speaking and interruptions during primary care office visits. *Fam Med.* 2001;33:528–532.

7 **In these recorded encounters:** Beckman HB, Frankel RM. The effect of physician behavior on collection of data. *Ann Intern Med.* 1984;101:692–696.

7 **In one study, over half of the patients interviewed:** Baker LH, O'Connell

D, Platt FW. What else? Setting the agenda for the clinical interview. *Annals Int Med.* 2005;143(10):776–771.

7 **In other studies doctor and patient disagreed:** Starfield B, Wray C, et al. The influence of patient-practitioner agreement on outcome of care. *Am J Public Health.* 1981;71:127–131. Burack RC, Carpenter RR. The predictive value of the presenting complaint. *J Fam Pract.* 1983;16:749–754.

7 **'If you ask questions':** Epstein RM, Street RL. Patient centered care for the 21st century: physicians' roles, health systems and patients' preferences. *ABIM.* 2008 Summer Forum, 'From Rhetoric to Reality: Achieving Patient Centered Care.'

7 **'you can never foretell':** Doyle AC. 'The Sign of Four,' *Sherlock Holmes: The Complete Novels and Stories*, vol. 1. NY: Bantam, 1986, p. 175.

7 **differences between the average and the individual:** Fosarelli P. Medicine, spirituality and patient care. *JAMA.* 2008;300(7):836–838.

8 **'What the patient brings to the process':** Platt F. Two collaborating artists produce a work of art: the medical interview. *Arch Int Med.* 2003;163:1131–1132.

8 **A visit to a doctor's office:** Forem J. Make the most of a doctor's visit. *Boston Globe,* September 19, 2005.

8 **In 1989, the average doctor's appointment:** Mechanic D, et al. Are patient office visits with physicians getting shorter? *N Engl J Med.* 2001;344(3):198–204.

8 **Studies suggest that getting a good history:** Stewart M, et al. The impact of patient-centered care on outcomes. *J Fam Pract.* 2000;49(9):796–804. Levinson W, et al. A study of patient clues and physician responses in primary care and surgical settings. *JAMA.* 2000;284:1021–1027.

8 **can even reduce visit time:** Mauksch LB, et al. Relationship, communication and efficiency in the medical encounter. *Arch Int Med.* 2008;168(13):1387–1395.

8 **patient satisfaction is higher:** Stewart M, et al. The impact of patient-centered care on outcomes. *J Fam Pract.* 2000;49(9):796–804.

12 **similar patterns in other patients:** Allen JH, et al. Cannabinoid

hyperemesis: cyclical hyperemesis in association with chronic cannabis abuse. *Gut.* 2004;52:1566–1570.

12 **Other case reports followed:** Allen JH, de Moore GM, et al. Cannabinoid hyperemesis: cyclical hyperemesis in association with chronic cannabis abuse. *Gut.* 2004;52;1566–1570. De Moore GM, Baker J, et al. Psychogenic vomiting complicated by marijuana abuse and spontaneous pneumonmediastinum. *Aust NZJ Psychiatry.* 1996;30:290–294. Roche E, Foster PN. Cannabinoid hyperemesis: not just a problem in Adelaide Hills. *Gut.* 2005;54:731.

14 **Studies have repeatedly shown:** Hill J. Effect of patient education on adherence to drug treatment for rheumatoid arthristis. *Ann Rheumatic Dis.* 2001;60:869–875. Kripalani S, et al. Interventions to enhance medication adherence in chronic disease. *Arch Int Med.* 2007;167(6):540–549.

14 **Patients who understand their illness:** Lin E HB, et al. Working with patients to enhance medication adherence. *Clin Diabetes.* 2008;26:17–19.

16 **A diagnosis of terminal cancer:** Cassell EJ. Diagnosing suffering: a perspective. *Annal Int Med.* 1999;131:531–534.

***Chapter 2: The Stories They Tell***

28 **Current thinking focuses on stories as the key:** Lucey CR. From problem lists to illness scripts: a new strategy to learn and teach professional thinking in small groups, a lecture given 1/14/03. Hunter KM. *Doctors' Stories: The Narrative Structure of Medical Knowledge.* Princeton, NJ: Princeton University Press, 1991, p. 17. Montgomery K. *How Doctors Think: Clinical Judgment and the Practice of Medicine.* New York: Oxford University Press 2006;45–53. Bowen JL. Educational strategies to promote clinical diagnostic reasoning. *N Engl J Med.* 2006;355:2217–25.

28 **These stories, what researchers now call illness scripts:** Schmidt HG, Rikers RMJP. How expertise develops in medicine: knowledge encapsulation and illness script formation. *Med Ed.* 2007;41:1133–39, Charlin B, et al. Scripts and clinical reasoning. *Med Ed.* 2007;41:1178–84.

29 **One of the ways doctors are taught to think about disease:** Mangruikar RS, et al. What is the role of the clinical 'pearl.' *Am J Med.* 2002;

113(7):617–24. Ioannidis JPA, Lau J. Uncontrolled pearls, controlled evidence, meta-analysis and the individual patient. *J Clin Epidemiolo.* 1998;51(8):709–11.

31 **Dr André Lemierre, a physician in Paris, first described this disease in 1936:** Lemierre A. On certain septicemias due to anaerobic organisms. *Lancet.* 1936;1:701–3. Centor RM. Should Lemierre's syndrome re-emergence change pharyngitis guidelines? Manuscript from author. Singhal A, Kerstein MD. Lemierre's syndrome. *Medscape.* 2001;94(9):886–87. http://www.medscape.com/viewarticle/410830.

32 **Like those presented to Fitzgerald:** Hunter KM. *Doctors' Stories: The Narrative Structure of Medical Knowledge.* Princeton, NJ: Princeton University Press, 1991.

34 **Anderson spent a year interviewing patients about their experiences in the health care setting:** Anderson A. *On the Other Side: African Americans Tell of Healing.* Louisville, KY: Westminster John Knox Press, 2001.

***Chapter 3: A Vanishing Art***

39 **A man in his fifties comes to an emergency room:** Jauhar S. The demise of the physical exam. *N Engl J Med.* 2006;354:548–551.

42 **most of the doctors holding a post-residency fellowship:** Mangione M, Nieman LZ, Kaye D, Gracely E. The teaching and practice of cardiac auscultation during internal medicine and cardiology training: a nationwide survey. *Annal Int Med.* 1993;119(1):46–54.

42 **If letter grades were being handed out:** Mangione S, Nieman LZ. Pulmonary auscultatory skills during training in internal medicine and family practice. *Am J Resp & Crit Care Med.* 1999;159(4 pt 1):1119–1124.

43 **Residents, their teacher-physicians:** Vukanovic-Criley JM, Criley S, et al. Competency in cardiac examination skills in medical students, trainees, physicians and faculty. *Arch Int Med.* 2006;166:610–616.

43 **Perhaps, Mangione suggests:** Mangione S. Teaching and practice of cardiac auscultation during internal medicine and cardiology training. *Ann Int Med.* 1993;119(1):47–54. Mangione S, Nieman L. Pulmonary auscultatory skills during training in internal medicine and family practice. *Am J*

*Resp Crit Care.* 1999;159:1119–1124. Mangione S, Duffy FD. The teaching of chest auscultation in primary care training: has anything changed in the 1990s. *Chest.* 2003;124(4):1430–1436.

44 **In 1980 the average length of stay:** Chassin MR. Variations in length of stay: their relationship to health outcomes. Report for the Office of Technical Assessment, US Congress, Washington, D.C., 1983.

44 **In a recent study done at Yale:** Private communication, John Moriarty, Associate Program Director, Yale Tradition, Internal Medicine Residency Program.

45 **In 1950 approximately 15,000 people died of rheumatic heart disease:** http://www.americanheart.org/presenter.jhtml?identifier=4712.

52 **In a study published in 2002:** McGreevy KM, et al. Clinical breast examination – practices among women undergoing screening mammography. *Radiology.* 2002;24:555–559.

***Chapter 4: What Only the Exam Can Show***

56 **The patient's story contained the diagnostic tip-off:** Hampton JR, et al. Contribution of history-taking, physical examination and laboratory evaluation to diagnosis and management of medical outpatients. *BMJ.* 1975;2(5969):486–489. Sandler G. The importance of the history in the medical clinic and the cost of unnecessary tests. *Am Heart J.* 1980;100(pt 1):928–931.

56 **When he couldn't find a good answer:** Reilly BM. Physical examination in the care of medical inpatients: an observational study. *Lancet.* 2003;362:1100–1105.

***Chapter 5: Seeing Is Believing***

84 **'If the patient's normal appearance is preserved':** From 'Prognosis' in *Hippocratic Writings*. NY: Penguin Books, 1983, p. 171.

84 **'By realising and announcing beforehand':** Ibid., p. 170.

88 **The decision to either admit the patient:** Mellors JW, Horwitz RI, et al. A simple index to identify occult bacterial infection in adults with acute unexplained fever. *Arch Int Med.* 1987;147(4):666–671.

90 **'I have trained myself':** Doyle AC. 'The Adventure of the Blanched Soldier,' *Sherlock Holmes: The Complete Novels and Stories*, vol. 1. NY: Bantam, 1986.

91 **But the most important trait they shared:** Several sources were used in researching this: Klauder JV. Sherlock Holmes as a dermatologist. *Arch Derm Syphilology.* 1953;68(4):363–377. Reed J. A medical perspective on the adventures of Sherlock Holmes. *Med Humanit.* 2001;27:76–81. Massey EW. Joseph Bell MD – Mr. Sherlock Holmes? *South Med J.* 1980;73(12):1635–1636. Scarlett EP. The old original: notes on Dr Joseph Bell whose personality and peculiar abilities suggested the creation of Sherlock Holmes. *Arch Int Med.* 1964;114:696–701. Conan Doyle dead from heart attack. *New York Times*, July 8, 1930. Wisser KM. The creation, reception and perpetuation of the Sherlock Holmes phenomenon 1887–1930. Master's thesis, University of North Carolina, Chapel Hill, 2000. Leibow E. *Dr Joe Bell: Model for Sherlock Holmes.* Bowling Green, OH: Bowling Green University Popular Press, 1982.

92 **'From close observation and deduction':** Leibow E. *Dr Joe Bell: Model for Sherlock Holmes.* Bowling Green, Ohio: Bowling Green University Popular Press, 1982.

94 **These students also improved:** Dolev JC, Friedlaender LK, Braverman IM. Use of fine art to enhance visual diagnostic skills. *JAMA.* 2001;286 (9):1020–1021.

95 **On a monitor I see six adults:** This video was conceived and produced by Daniel J. Simons, Associate Professor, University of Illinois, Visual Cognition Lab. Screen it for your friends by going to http://viscog.beckman.uivc.edu/djs_lab/index.html.

96 **So did more than half:** Simons DJ, Chabris CF. Gorillas in our midst: sustained inattention blindness for dynamic events. *Perception.* 1999;28:1059–1074.

97 **Researchers call this phenomenon:** Chun MM, Marois R. The dark side of visual attention. *Curr Op Neurobio.* 2002;12:184–189. Most SB, Scholl BJ, Clifford ER, Simons DJ. What you see is what you set: sustained inattentional blindness and the capture of awareness. *Psych Rev.* 2005;112(1):217–242.

106 **Participants in the study were shown two pictures:** Kelley TA, et al. Effect of scene inversion on change detection of targets matched for visual salience. *Journal of Vision.* 2003;2:1–5.

***Chapter 6: The Healing Touch***

109 **'It is the business of the physician':** Adams CD, ed. *The Genuine Works of Hippocrates.* NY: Dover, 1868, from *The Digital Hippocrates,* http://www.chlt.org/sandbox/dh/Adams/page.160.a.php.

112 **The third that didn't have the CT scan:** Musunuru S, Chen H, et al. Computed tomography in the diagnosis of acute appendicitis: definitive or detrimental. *J Gastrointest Surg.* 2007;11:1417–1422.

125 **Mammographers agreed with one another:** Elmore JG, Wells CK, et al. Variations in radiologists' interpretation of mammograms. *N Engl J Med.* 1994;331:1493–1499.

***Chapter 7: The Heart of the Matter***

133 **'tip of the iceberg':** Salvatore Mangione, personal communication.

136 **'on account of the great degree of fatness':** Nuland SB. *Doctors: The Biography of Medicine.* NY: Vintage Books, 1995, p. 220.

136 **'I recalled a well known acoustic phenomenon':** Duffin J. *To See with a Better Eye: The Life of RTH Laennec.* Princeton, NJ: Princeton University Press, 1998, p. 122.

137 **They argued that diseases could be classified:** Ibid., p. 26.

139 **It was usually attributed to heart failure:** I usually explain heart failure to my patients using a scene from *I Love Lucy.* In this episode, Lucy and her friend Ethel take jobs in a candy factory, wrapping candies as they move by on a conveyer belt. Initially they are able to keep up and all the candies end up neatly wrapped. As the conveyer belt picks up speed, more candies are delivered, and it becomes harder and harder to keep up. Before long the two are overwhelmed and candies end up everywhere – in their pockets, in their blouses, on the floor. What happens to Ethel and Lucy is analogous to what happens to the heart – with even the slightest physical stress, the weakened organ is overwhelmed by the amount of blood being brought in

and, like the overflowing candies, the extra fluid backs up, ending up just about everywhere.

139 **the now classic finding of emphysema:** Duffin J. *To See with a Better Eye: The Life of RTH Laennec.* Princeton, NJ: Princeton University Press, 1998, pp. 157–58.

139 **'oppression and palpitations':** Major RH. *Classic Descriptions of Disease.* Springfield, IL: Charles C Thomas Publisher, 1932, pp. 371–372.

145 **In one large multispecialty group in Boston:** Blanchard GP. Is listening through a stethoscope a dying art? *Boston Globe,* May 25, 2004.

146 **five cardiologists were pitted against echocardiography:** Jaffe WM, et al. Clinical evaluation versus Doppler echocardiogram in the quantitative assessment of valvular heart disease. *Circulation.* 1988;78:267–275.

146 **In a study done by Christine Attenhofer:** Attenhofer Jost CH, Turina J, Mayer K, Seifert B, Amann FW, Buechi M, et al. Echocardiography in the evaluation of systolic murmurs of unknown cause. *Am J Med.* 2000;108:614–620.

146 **One study done of emergency room physicians:** Reichlin S, et al. Initial clinical evaluation of cardiac systolic murmurs in the ED by non-cardiologist. *Am J Emerg Med.* 2004;22:71–75.

147 **Several studies have been done evaluating programmes:** Smith CA, et al. Teaching cardiac examination skills: a controlled trial of two methods. *J Gen Int Med.* 2006;21(1):1–6. Barrett MJ. Mastering cardiac murmurs: the power of repetition. *Chest.* 2004;126:470–475. Favrat B, et al. Teaching cardiac auscultation to trainees in internal medicine and family practice: does it work? *BMC Med Ed.* 2004;4:5. http://www.biomedcentral.com/1472-6920/4/5.

161 **'Direct observation of trainees':** Holmboe ES, Hawkins RE. Evaluating the clinical competence of residents in internal medicine: a review. *Ann Int Med.* 1998;129:42–48.

162 **A study published recently shows how inadequate:** Hicks CM, et al. Procedural experience and comfort level in internal medicine trainees. *J Gen Intern Med.* 2000;15:716–722.

***Chapter 8: Testing Troubles***

171 **Furthermore, there is plenty of solid evidence:** Klempner MS, et al. Two controlled trials of antibiotic treatment in patients with persistent symptoms and a history of Lyme disease. *N Engl J Med.* 2001;345:85–92.

172 **Finally, in October 1975:** Clark E. Lyme disease: one woman's journey into tick country. http://www.yankeemagazine.com/issues/2007-07/features/lymecountry.

173 **Comparing the location of Steere's mystery cases:** Steere AC. David France, scientist at work. *New York Times,* May 4, 1999.

173 **In order to transmit the infection:** Steere AC, et al. The emergence of Lyme disease. *J Clin Invest.* 2004;113(8):1093–1101.

173 **some studies suggest that the most common presentation:** Tibbles CD, et al. Does this patient have erythema migrans. *JAMA.* 2007;297:2617–2627.

178 **Post–Lyme Disease syndrome:** Steere AC, et al. Association of chronic Lyme arthritis with HLA-DR4 and HLA-DR2 alleles. *N Engl J Med.* 1990;323:219–223.

178 **They recruited one hundred residents:** Shadick NA, Phillips CB, Logigian EL, Steere AC, Kaplan RF, Berardi VP, et al. The long-term clinical outcomes of Lyme disease. A population-based retrospective cohort study. *Ann Int Med.* 1994;121:560–567.

178 **Other studies too have found:** Cairn V, Godwin J. Post-Lyme borreliosis syndrome: a meta-analysis of reported symptoms. *Int J Epi.* 2005;34:1340–1345.

179 **Researchers at Tufts Medical Center:** Klempner MS, et al. Two controlled trials of antibiotic treatment in patients with persistent symptoms and a history of Lyme disease. *N Engl J Med.* 2001;345:85–92.

179 **Two other rigorous trials:** Krupps LB, et al. Study and treatment of post Lyme disease. *Neurology.* 2003;60:1923–1930. Fallon BA. A randomized, placebo-controlled trial of repeated IV antibiotic therapy for Lyme encephalopathy. *Neurology.* 2008;70:992–1003.

181 **They don't trust either physical exams:** The International Lyme and Associated Diseases Society Evidence-based guidelines for the management

of Lyme disease, published November 2006, p. 7, http://www.ilads.org/guidelines.html, accessed December 31, 2007.
181 **In fact, when used as recommended:** Tugwell P, et al. Laboratory evaluation in the diagnosis of Lyme disease. *Ann Int Med.* 1997;127(12):1109–1123.
181 **These are some of the most common symptoms:** Fletcher K. Ten most common health complaints. *Forbes,* July 15, 2003. http://www.forbes.com/2003/07/15/cx_kf_0715health.html.

***Chapter 9: Sick Thinking***
197 **Diagnostic errors are the second leading cause:** Bartlett EE. Physicians' cognitive errors and their liability consequences. *J Healthcare Risk Manage.* 1998(fall):62–69.
197 **And a recent study of autopsy findings:** Tai DYH, El-Bilbeisi H, Tewari S, Mascha EJ, Wiedermann HP, Arroliga AC. A study of consecutive autopsies in a medical ICU: a comparison of clinical cause of death and autopsy diagnosis. *Chest.* 2001;119:530–536.
198 **One survey showed that over one third of patients:** Berner ES, et al. Overconfidence as a cause of diagnostic error in medicine. *Am J Med.* 2008;121(5A):S2–S23.
200 **Faulty synthesis . . . by comparison, played a role:** Errors of inadequate data collection are probably underrepresented in this sample because it was based on chart review. If something was missed, it won't be in the chart. To pick up this kind of error requires access to the patient at the time of the diagnosis.
201 **'Thinking stops when a diagnosis is made':** Croskerry P. The importance of cognitive errors in diagnosis and strategies to minimize them. *Acad Med.* 2003;78(8):1–6.
201 **'process of matching':** Croskerry P. Overconfidence in clinical decision making. *Am J Med.* 2008;121(5A):S24–S29.
201 **'the power of thin slicing':** Gladwell M. *Blink.* NY: Little, Brown, 2005. http://www.gladwell.com/blink/.
203 **'The trick lies in matching':** Croskerry P. The theory and practice of clinical decision-making. *Can J Anesth.* 2005;52(6):R1–R8.

206 **black men are significantly more likely:** http://www.cdc.gov/cancer/prostate/statistics/race.htm, accessed May 1, 2008.
207 **'despite their "objective" medical training':** McKinlay JB, Potter DA, Feldman HA. Non-medical influences on medical decision-making. *Soc Sci Med.* 1996;42(5):769–776.
207 **And even those factors:** Ibid.
207 **One of the many careful experiments:** Arber S, McKinlay J, Adams A, Marceau L, Link C, O'Donnell A. Patient characteristics and inequalities in doctors' diagnostic and management strategies relating to CHD: a video-simulation experiment. *Soc Sci Med.* 2006;62(1):103–115.
214 **In the 1930s:** Gawande A. 'The Checklist.' *The New Yorker,* 12/10/07, http://www.newyorker.com/reporting/2007/12/10/071210fa_fact_gawande.
214 **These basic steps:** Wachter RE. *Understanding Patient Safety.* New York: McGraw-Hill Medical, 2008, p. 23.
215 **surgical safety checklist:** Haynes AB et al. A surgical safety checklist to reduce morbidity and mortality in a global population. *N Engl J Med.* 2009; 360, pp. 491–499.
215 **checklist before certain procedures in the ICU:** Pronovost Petal. An intervention to decrease catheter-related bloodstream infections in the ICU. *N Engl J Med.* 2006;355, pp. 2725–2732.

***Chapter 10: Digital Diagnosis***

216 **In a 1976 article, a group of doctors:** Pauker SG, Gorry GA, Kassirer JP, Schwartz WB. Towards the simulation of clinical cognition taking a present illness by computer. *Am J Med.* 1976;60:981–996.
221 **In 1994 she and a group of thirteen other physicians:** Berner ES, Webster GD, Shugerman AA, et al. Performance of four computer-based diagnostic systems. *N Engl J Med.* 1994;330:1792–1796.
229 **In order to measure how well the program can perform:** Leonhardt D. Why doctors so often get it wrong. *New York Times,* February 22, 2006.
229 **Mark Graber and a colleague:** Graber MI, Matthew A. Performance of a web-based clinical diagnosis support system for internists. *J Gen Int Med.* 2008;23(supp 1):37–40.

231 **According to a 2005 survey done by the Pew Center:** Boone S. Computer users can catch the health bug on line. *McClatchy-Tribune Regional News–The Walton Sun*, November 17, 2007.

234 **Finally, the visiting professor asked the fellow:** Greenwald R. . . . And a diagnostic test was performed. *N Engl J Med* (letter). 2005;353:2089–2090.

235 **This story and their own experiences with patients:** Tang H, Hwee Kwoon Ng J. Googling for a diagnosis – use of Google as a diagnostic aid: Internet based study. *BMJ.* 2006;333;1143–1145.

236 **Even the august *New England Journal of Medicine*:** Fan E et al. A gut feeling. *N Eng J Med.* 2008;359:75–80.

***Afterword: The Final Diagnosis***

241 **Medicine's first toehold:** Much of this comes from Roy Porter's wonderful history of medicine, *The Greatest Benefit to Mankind.* NY: Norton, 1999; as well as from Jacalyn Duffin's biography of René Laennee, *To See with a Better Eye*. Princeton, NJ: Princeton University Press, 1998.

242 **These days, patients who die in a hospital:** David Dobb wrote a terrific piece about autopsy, 'Buried Answers,' for the *New York Times* magazine, April 24, 2005.

243 **Small residency programmes objected to the ever growing cost:** From Accreditation Council for Graduate Medical Education, a private nonprofit council that evaluates and accredits medical residency programmes in the United States, personal communication.

243 **Certainly a doctor's ability to make an accurate diagnosis:** A report from the Agency for Healthcare Research and Quality, written by Washington AE and McDonald KM, The autopsy as an outcome and performance measure (Evidence Report/Technology Assessment 58, October 2002), provided much of the information on the modern history of the autopsy.

# INDEX